WELL SERVICING AND WORKOVER

Petroleum Geology and Reservoirs

LESSON 2

2nd edition

Published by

PETROLEUM EXTENSION SERVICE
The University of Texas at Austin
Austin, Texas

in cooperation with

ASSOCIATION OF OILWELL SERVICING CONTRACTORS
Dallas, Texas
1992

Library of Congress Cataloging-in-Publication Data

Morris, Jeff

Petroleum geology and reservoirs/by Jeff Morris. — 2nd ed.

p. cm.— (Lessons in well servicing and workover; lesson 1

ISBN 0-88698-156-5

1. Petroleum—Geology—Programmed instruction. 2. Natural gas—Geology—Programmed instruction. 3. Oilfields—Programmed instruction. 4. Gas reservoirs—Programmed instruction. I. Title. II. Series

TN870.5.M585 1992

622'.3382—dc20 90-22077

CIP

First edition published 1971. Second edition 1992
Fourth Impression 2007
Printed in the United States of America

Catalog No. 3.70220
ISBN 0-88698-156-5

CONTENTS

FIGURES

TABLES

FOREWORD

The Human Resources Committee of the Association of Oilwell Servicing Contractors (AOSC) is the sponsor of twelve lessons on well servicing and workover. These lessons are designed to be used as home study lessons—that is, students or trainees can read the lessons at their own pace. Questions are provided to help students determine whether they have learned the material covered in each lesson. These lessons can be covered in roughly two years of home study time, although many will find it possible to complete the study in a shorter period.

Most of the learning by those employed in the servicing and workover industry takes place on the job, with aid from supervisors and associates in the field. These lessons are primarily designed to help new employees find out what the work is about and to better understand their part in the operation. Further, by knowing how to do a job properly, employees should also do it safely.

The Human Relations Committee of the AOSC, chaired by Harold Foster, Freberg and Company of Wyoming, Inc., should be credited for its endorsement and encouragement in the preparation of these lessons.

Lesson 2 describes elementary geology and reservoir information as they pertain to petroleum. The first edition was written in 1971 by W. E. Boyd, director of Petroleum Extension Service (PETEX) at that time. The material in this edition has, however, been updated extensively and reorganized and new material has been added.

While this lesson could not have been completed without the cooperation and assistance of the many well servicing and workover contractors who are members of the AOSC, PETEX is solely responsible for its content. Also, while every effort was made to ensure that accurate and up-to-date information is provided, this manual is intended for educational purposes only. Nothing in it is approval or disapproval of any product or practice.

Ron Baker, director
Petroleum Extension Service
Division of Continuing Education
The University of Texas at Austin

Austin, Texas
1992

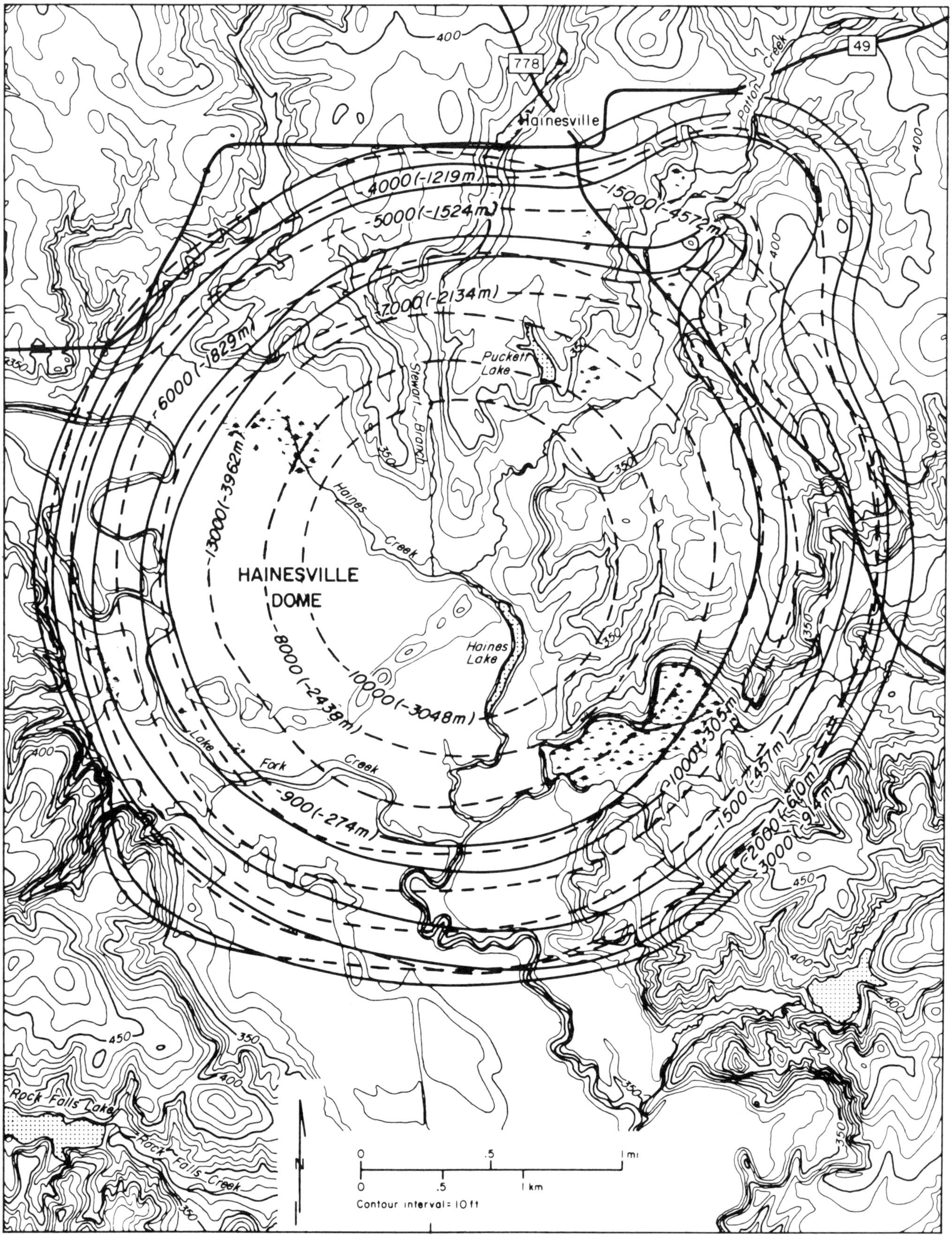
778
49
Hainesville
Patton Creek
4000(-1219m)
5000(-1524m)
15000(-4572m)
7000(-2134m)
6000(-1829m)
Puckett Lake
Stewart Branch
13000(-3962m)
Haines Creek
HAINESVILLE DOME
Haines Lake
8000(-2438m)
10000(-3048m)
Lake Fork Creek
900(-274m)
1000(-305m)
1500(-457m)
2000(-610m)
3000(-914m)
Rock Falls Lake
Rock Falls Creek
0 .5 1 mi
0 .5 1 km
Contour interval = 10 ft
N

CHAPTER 1
BASIC GEOLOGIC CONCEPTS

Introduction

In the petroleum industry, which finds and recovers oil and gas from deep within the earth's crust, geology is fundamental. Petroleum occurs mostly in isolated, hard-to-find accumulations. The scientific study of the earth's history and its life, especially as recorded in the rocks of the crust, reduces the risk of drilling dry holes and lowers the cost of production by helping determine the most efficient way of drilling a well. A knowledge of geology increases the total supply of petroleum by helping recover more of the resource in place.

Petroleum geologists are most concerned with rocks formed in the earth's surface by processes closely associated with both climate and life. The way these rocks are created and changed, as well as how oil and gas form and accumulate in them, are the principal concerns of the petroleum geologist. For a thorough understanding of these processes, it is necessary to look back in time—first, to the beginning of the modern science of geology; then, to the beginning of the earth itself.

Uniformitarianism

Ancient geologists believed that the earth had been created all at once, complete with all its mountains, canyons, and oceans, in a single great cataclysm. In the 1700s, though, scientists began to understand that familiar natural processes, such as the accumulation and erosion of sediment, and "minor" cataclysms, such as earthquakes and volcanic eruptions, could account for all the features of the earth's crust—given enough time. Thus the doctrine of *catastrophism* was eventually supplanted by the theory of *gradualism* or *uniformitarianism*—which holds, as Scottish geologist James Hutton put it two centuries ago, that "the present is the key to the past."

This concept of gradual change is central to modern geology. Today's geologists know that the Grand Canyon is the work of a powerful erosive agent, the Colorado River, over some 10 million years (fig. 1); that the Himalayas and the Sierra Nevada are growing loftier by a fraction of an inch each year, and have been doing so for millions of years; that Africa and America are moving away from each other about as fast as a fingernail grows.

Figure 1. Grand Canyon

Geologic Time

Geologists now obtain close estimates of the age of rocks by measuring their radioactivity. Naturally occurring radioactive elements, such as uranium, change at a measurable rate into other elements, such as lead. By measuring the proportions of different forms of lead, scientists can tell about how much time has passed since a rock was formed. Using such methods, geologists have radically changed our ideas about the age of the planet.

Even the 10 million years that it took to carve the Grand Canyon is but the most recent moment of geologic history. The earth was formed about 4.6 *billion* years ago when frozen particles and gases circling a new yellow star were brought together by mutual gravitational attraction. Heated by compression and radioactivity, this material formed a molten sphere. The heaviest components, mostly iron and nickel, sank to the center and became the earth's *core*. Lighter minerals formed a thick, molten *mantle*, while minerals rich in aluminum, silicon, magnesium, and other light elements cooled and solidified into a thin, rocky *crust* (fig. 2).

The surface of the young planet was an inhospitable place. Molten rock *(magma)* erupted everywhere through fissures and volcanoes, expelling the gases and water vapor that formed the early, oxygenless atmosphere. As the surface cooled, rain condensed and fell in torrents, and the first oceans began to form.

The earth was devoid of life for perhaps its first billion years. Eventually, out of a mixture of complex carbon-chain chemicals, the first self-replicating molecules appeared in the ocean, perhaps in the muck of some shallow lagoon. Over millions of years these primitive organisms grew more complex and varied, first as single-celled bacterialike forms, later as microscopic protozoa and algae. Some grew in the form of colonies, which over further millions of years evolved into more complex organisms. As photosynthetic single-celled plants, which used carbon dioxide and gave off oxygen, became more abundant, their waste oxygen became a major constituent of the atmosphere.

Few traces of this early life survive, however. Although plant remains and impressions of primitive organisms can be found, it was about 4 billion years before animal life became abundant enough (and developed body parts durable enough) to leave significant numbers of fossils. This early, fossil-poor period, comprising most of the time since the earth formed, is commonly known as the Precambrian era (fig. 3).

The last 600 million years of earth's history comprise the time of abundant life. The first fish appeared about 500 million years ago in the early Paleozoic era, followed by the first land plants, amphibians, and reptiles. The Mesozoic (220 to 65 million years ago) was the era of the dinosaurs, early mammals, and primitive birds. And the Cenozoic era embraces the time from the extinction of dinosaurs through the recent ice ages to the present.

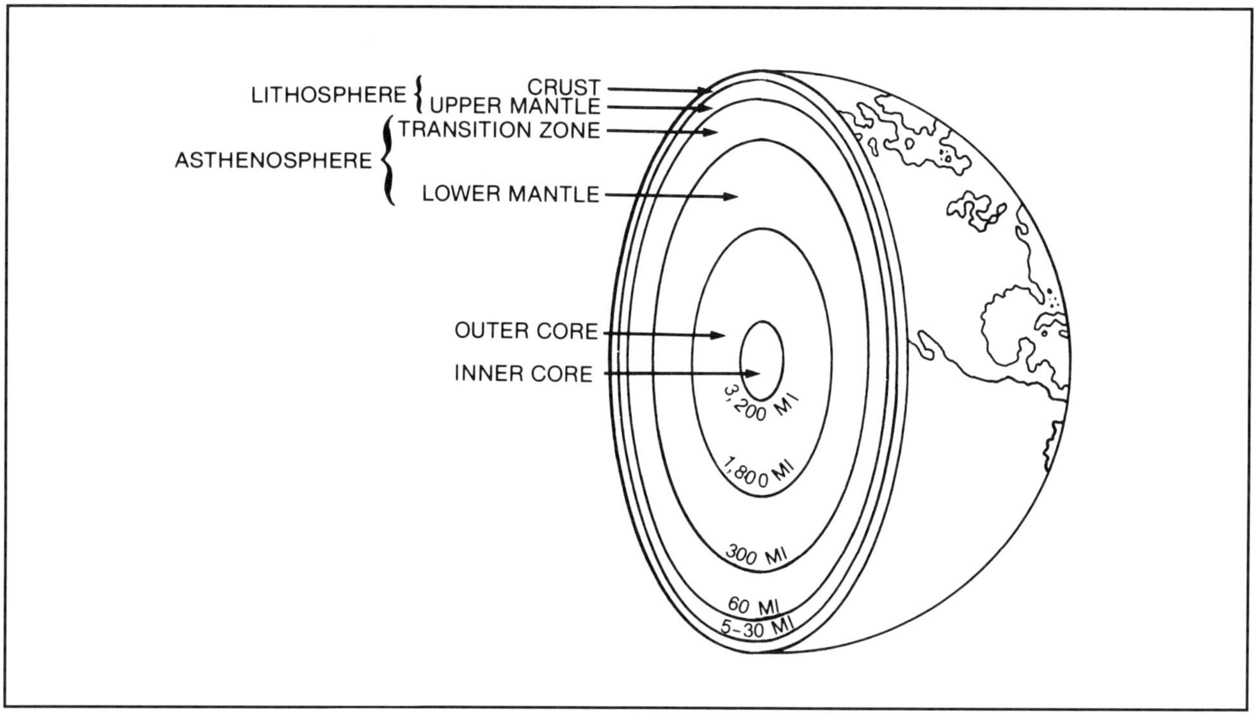

Figure 2. Cutaway view of Earth

ERA	PERIOD	EPOCH	AGE (MM. Yrs)	EVENTS
CENOZOIC	QUATERNARY	HOLOCENE	0.01	End of most recent ice advance in North America
		PLEISTOCENE		Recent ice ages
			1	
	TERTIARY	PLIOCENE		Colorado River begins cutting Grand Canyon
		MIOCENE		India collides with Asia
		OLIGOCENE		
		EOCENE		
		PALEOCENE		Rocky Mountains formed Mammals proliferate
			65	Dinosaurs extinct
MESOZOIC	CRETACEOUS			First flowering plants
			135	
	JURASSIC			First birds
			190	
	TRIASSIC			Atlantic rift opens First mammals First dinosaurs
			220	
PALEOZOIC	PERMIAN			
			280	Glaciation of southern Gondwanaland
	PENNSYLVANIAN			First reptiles
			325	
	MISSISSIPPIAN			Africa collides with North America (Appalachians form)
			350	
	DEVONIAN			First land animals (amphibians)
			400	First rooted land plants
	SILURIAN			
			430	Glaciation
	ORDOVICIAN			First vertebrates (fish)
			500	
	CAMBRIAN			
			600	
PRECAMBRIAN				First complex animals (worms) First green algae
			2000	World iron ore deposits forming
			3000	Oldest evidence of life
			3800	Oldest known rocks
			4600	Formation of planets, solar system

Figure 3. Geologic time

Plate Tectonics

The young earth's molten surface was in constant motion, like the lava in an active volcano today. As a solid crust began to form, it was carried about on the surface by the moving magma beneath. Although this crust has grown thicker and stronger over time, it is still in motion atop the moving mantle.

The crust is divided by a worldwide system of faults, trenches, and midocean ridges into six major plates and many minor plates that fit together like the pieces of a jigsaw puzzle (fig. 4). These plates, however, move and change shape. In some places, they slide past one another; in others, they collide or move apart. The theory that explains how these processes work to shape the crust is called *plate tectonics*.

The earth's surface consists of two kinds of crust (fig. 5). Oceanic crust is thin (about 5 to 7 miles) and dense. The rock that makes up the continents, however, is thick (10 to 30 miles) and relatively light. A continent rises high above the surrounding oceanic crust and extends deeper into the mantle—like an iceberg in a frozen-over sea.

The continental heights are gradually worn away, mostly by the persistent force of running water. Particles of rock are carried to lower elevations and eventually into the sea, where they are deposited in thick sedimentary beds just offshore (fig. 6). Cemented together by minerals in the water and by the pressure of more sediments deposited on top of them, these beds are transformed into rock.

Sometimes a plate splits and begins moving apart. This is the way ocean basins are formed. Figure 7 shows a rift forming in the middle of a continent. As the two parts of the continent pull away from each other, magma rises from the mantle and solidifies in the gap, forming a *midocean ridge*. New crust, thinner than the continents but denser, spreads outward between the two "daughter" continents.

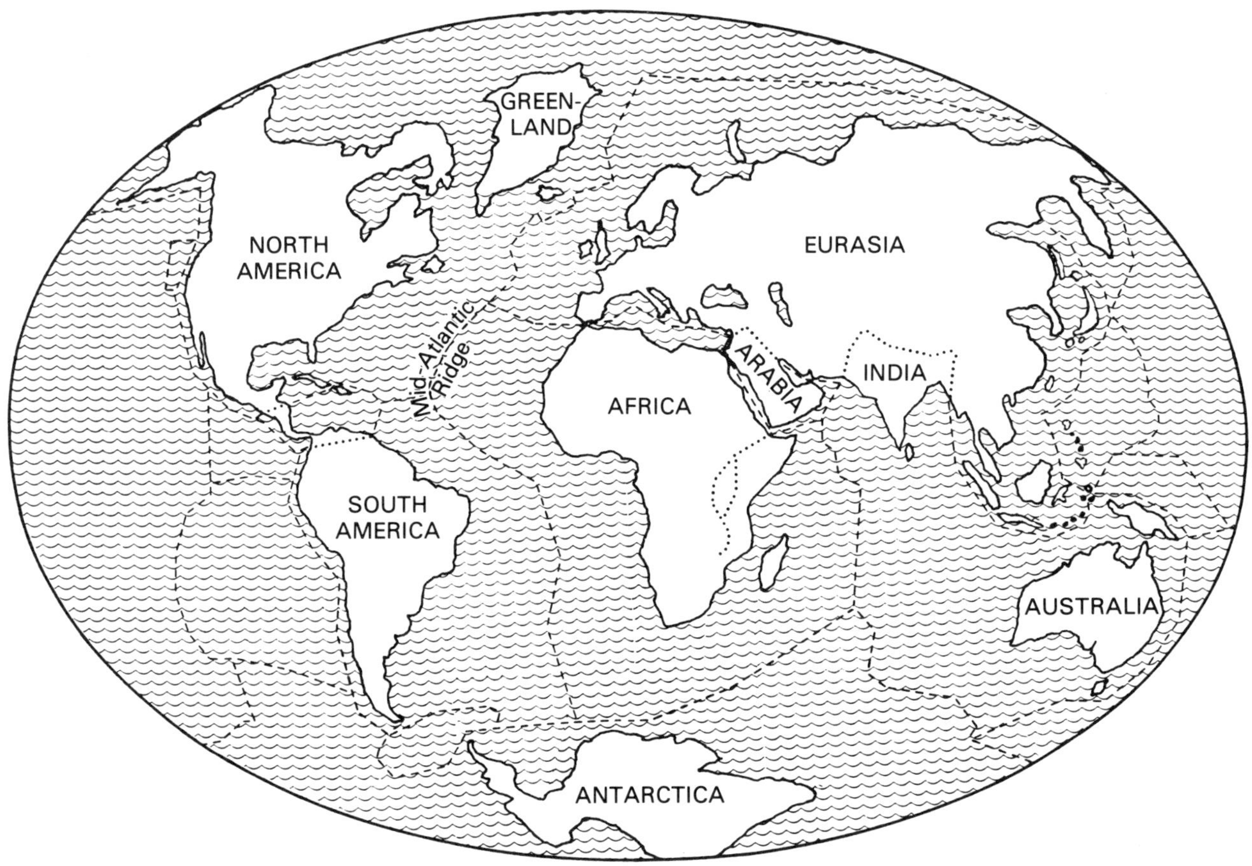

Figure 4. Major tectonic features of the earth's crust

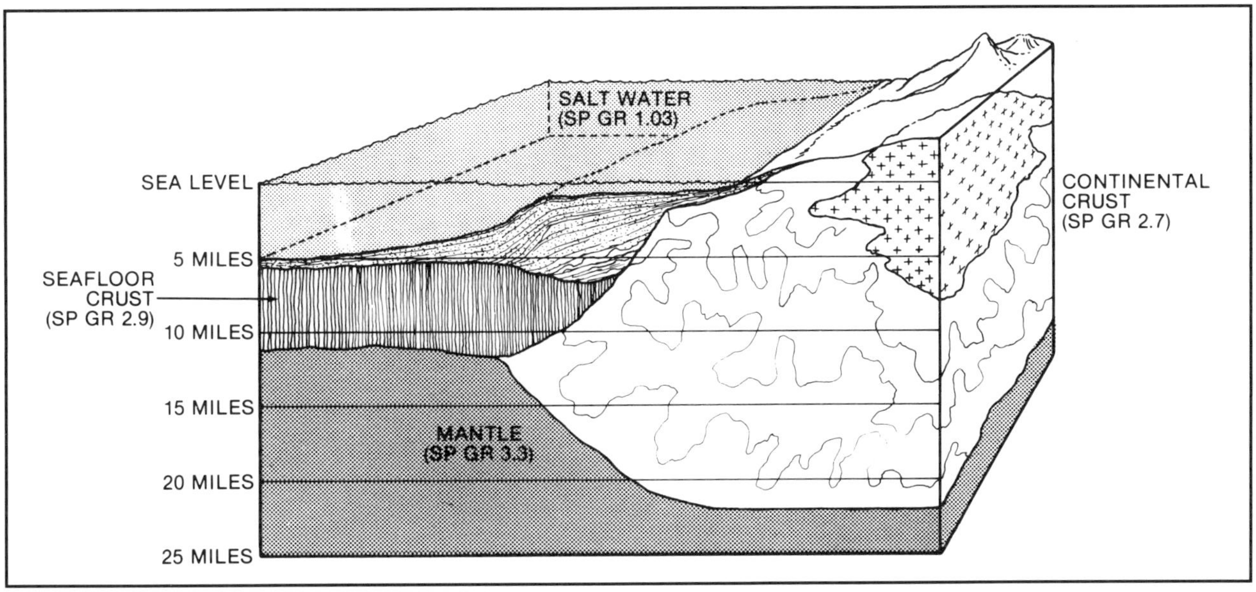

Figure 5. Relative thickness and specific gravity of seafloor and continental crust

The Atlantic Ocean was born in just this way about 200 million years ago when North and South America split away from Europe and Africa.

Where plates meet head on, several things can happen. If oceanic crust meets oceanic crust, one plate is *subducted*—that is, it slips beneath the edge of the other plate and descends into the mantle, forming a trench in the ocean floor (fig. 8). The descending plate is melted by the hot mantle in the subduction zone. Some of its minerals melt at lower

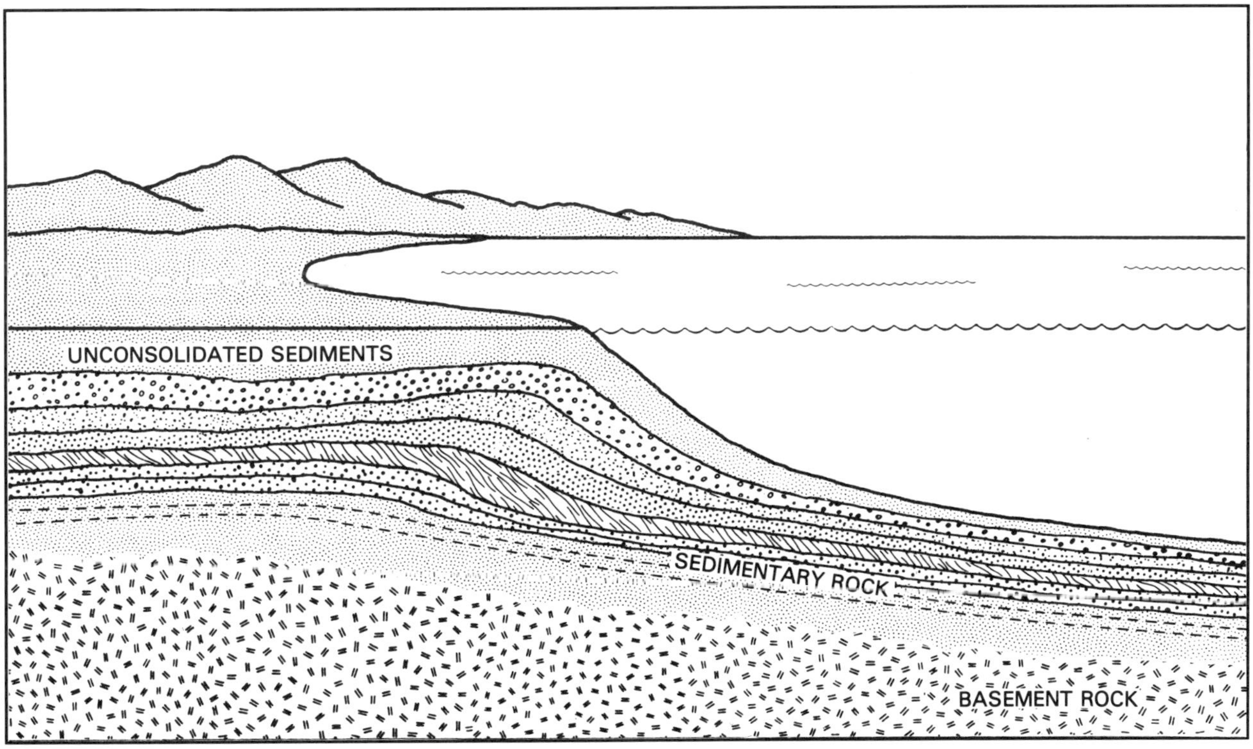

Figure 6. Burial and consolidation of sediments into rock

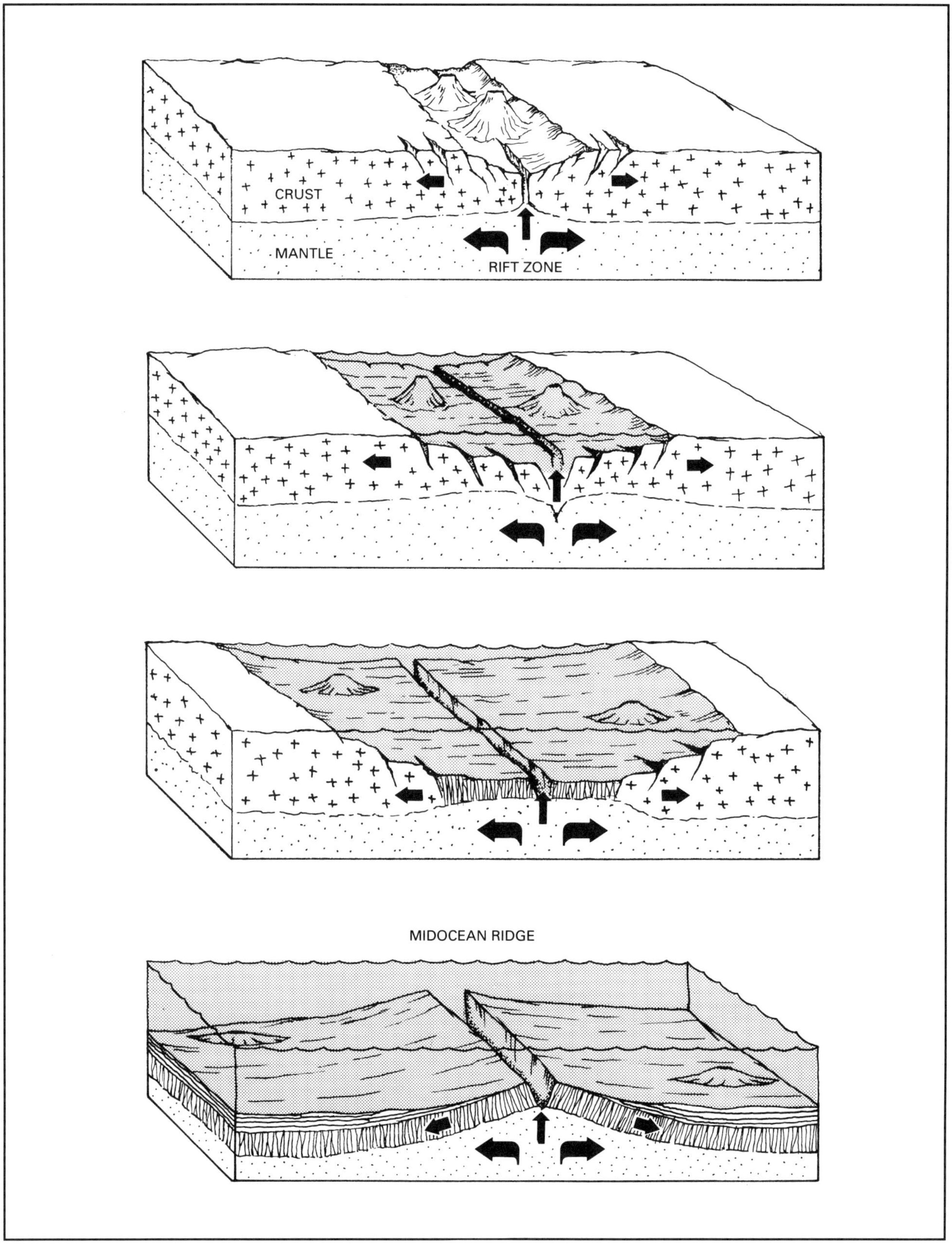

Figure 7. Rifting of continental crust

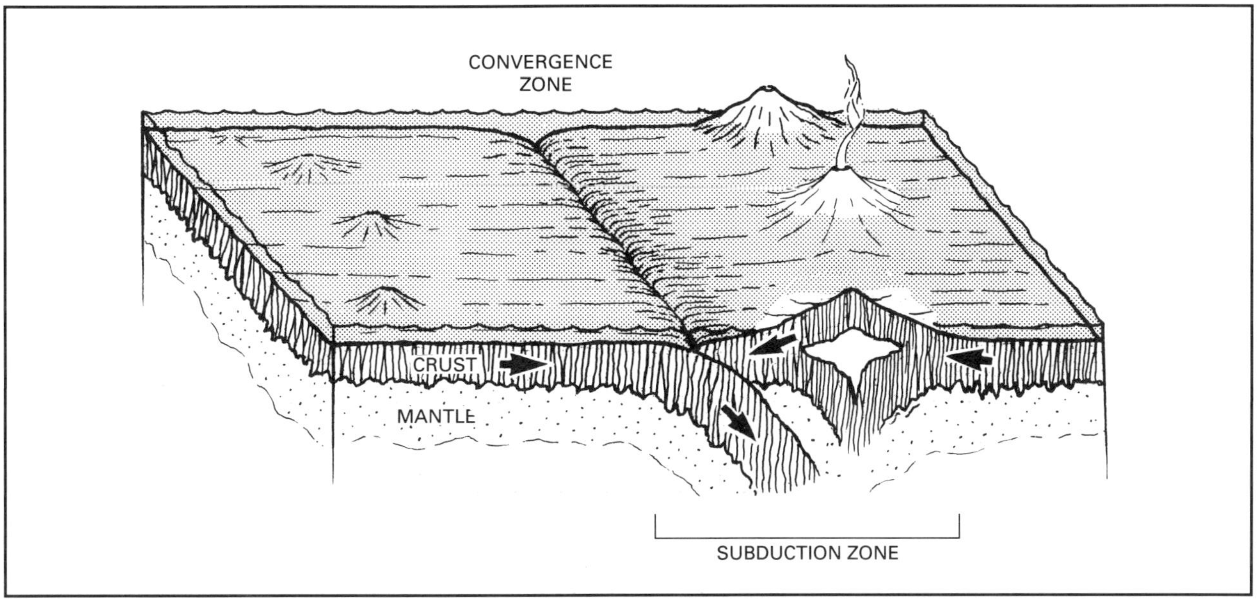

Figure 8. Subduction of oceanic crust

temperatures than others and rise through the crust as magma, which may either cool and solidify within the crust, forming igneous rock such as granite, or reach the surface as volcanic lava.

If one of the converging plates is made up of continental crust, it overrides the heavier oceanic plate, which bends downward in a trench along the continental margin. When this happens, magma from the descending plate may erupt in continental volcanoes like Mount St. Helens. If both of the plates are continental, the collision buckles and folds the rocks—including the sedimentary rocks at the edges of the continents—into great mountain ranges like the Himalayas.

Rocks and Minerals

In discussing the components of the earth's crust, it is important to distinguish between rocks and minerals. A *mineral* is a naturally occurring crystalline substance of a definite range of chemical composition. A *rock* is a mixture of minerals, usually in the form of grains that may be easily visible or microscopic. The most common rock minerals are *silicates*—crystalline compounds composed largely of silicon in chemical combination with aluminum, magnesium, oxygen, and other common elements.

Igneous rocks are those that cool and solidify from a molten state. They are classified by chemical composition and grain size. These characteristics, in turn, depend on the elements present in the magma and on how long they cool—the longer the cooling time, the larger the crystals.

Rocks that are exposed at the surface of the earth are subject to *weathering* by climatic agents, especially water. Water breaks down solid rock by changing it chemically, by dissolving some of its minerals, by supporting the growth of plants and animals that grow on and around rock, and by freezing and expanding to wedge the rock apart. Running water then carries fragments of rock and soil to sedimentary basins—low places where sediments can accumulate, sometimes to a depth of several miles. The weight of the accumulating sediments compresses and bonds the deeper beds into layers of sedimentary rock.

Any type of rock that is buried deeply enough or otherwise subjected to great pressure, stress, or heat can become transformed both chemically and physically into another kind of rock: *metamorphic* rock. For instance, *shale,* a crumbly sedimentary rock made of clay, can be changed by heat and pressure into hard metamorphic *slate.* Slate, or any other rock, can in turn be heated until it melts and then cooled into fresh igneous rock, or it can be broken down by weathering so that it contributes to the formation of new sedimentary rock. The principles involved in the transformation of one type of rock into another are illustrated by the *rock cycle* (fig. 9).

Two of the most important characteristics of sedimentary rocks, attributes that are rarely found in igneous and metamorphic rock, are their porosity and permeability. *Porosity* is the amount of empty space present within the rock; it is usually expressed as a percentage of total rock volume. *Permeability* is a measure of the ease with which a fluid flows through the connecting pore spaces of a rock; the more connections between pores, the higher the permeability. Porosity and permeability are of supreme importance to the geologist in determining whether a body of rock can contain petroleum and whether that petroleum can be extracted and brought to the surface.

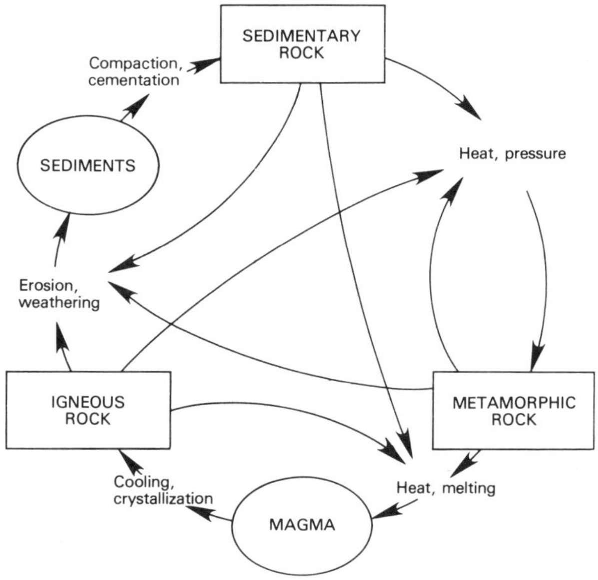

Figure 9. The rock cycle

REVIEW QUESTIONS

Chapter 1

1. The theory that explains how sections of the earth's crust move and interact is called plate techtonics.

2. Geology is the science of Earth's history.

C 3. The earth is estimated to be ____________ years old.
   - A. 10 million
   - B. 46 million
   - C. 4.6 billion
   - D. 100 billion

B 4. Which statement accurately describes uniformitarianism?
   - A. The Grand Canyon is the result of a major catastrophe about 10 million years ago.
   - B. Geologic changes are the result of processes operating over long periods of time.
   - C. All mountain ranges are the same age.
   - D. Mineral grains found in igneous rock are of uniform size.

A 5. New oceanic crust is created—
   - A. along midocean ridges.
   - B. in deep-sea trenches.
   - C. in mountain ranges.
   - D. beneath stream channels.

For questions 6–10, select the correct era from the list on the right. You may use an answer more than once.

D	6. The longest era	A. Cenozoic
C	7. Era when the first fish appeared	B. Mesozoic
D	8. Era with the fewest fossils	C. Paleozoic
A	9. The last 65 million years	D. Precambrian
C	10. Era when the first land plants appeared	

11. The three basic types of rock are sedimentary, igneous, and metamorphic.

T 12. (T/F) The faster magma cools, the smaller the crystals formed.

13. When oceanic crust and continental crust converge in a subduction zone, continental crust overrides oceanic crust.

14. The layer beneath the earth's crust is called the upper mantle.

D 15. The most abundant rock minerals are those containing—
   - A. carbon.
   - B. platinum.
   - C. uranium.
   - D. silicon.

CHAPTER 2
SEDIMENTARY ROCKS

Introduction

For the petroleum geologist, sedimentary rock is the most interesting type of rock. Some sedimentary rock formations are porous enough to hold great quantities of oil and gas; others contain high proportions of the organic matter from which, under certain conditions, hydrocarbons are generated.

Sedimentary Processes

Sedimentary rock is rock made up of fragments or chemical compounds from previously existing rocks or organisms. Carried by flowing water, ice, or air in response to the force of gravity, sediment accumulates in upland basins and along the edges of the continents. The depth of an accumulation can reach several miles (fig. 6). Deeply buried sediments are transformed into hardened rock by a set of processes called, collectively, *lithification*. The processes that alter the rock itself, either during or after its formation, are called *diagenesis*.

Compaction and *cementation* are two of the principal processes that change sediments into rock. As successive layers of water-saturated sediment accumulate, the deeper layers are compacted by the weight of overlying beds. The individual grains are forced into closer contact and, in some cases, are deformed. Minerals dissolved in the water—commonly, calcite (calcium carbonate, $CaCO_3$)—form a solid material that cements the grains together (fig. 10). Much of the water is squeezed out as the sediment is transformed into rock, but some becomes trapped in the pores as *connate,* or *interstitial,* water. Rock formed from sediments deposited by water almost always contains interstitial water.

Close study of sedimentary rock reveals the conditions under which it was formed. One set of conditions includes the events that occur beneath the surface during lithification and diagenesis—compaction, cementation, and chemical alteration by groundwater. The natural conditions that most influence the character of sedimentary rock are, however, those that occur at the earth's surface, where the solid earth is in contact with the fluids of the atmosphere and the oceans and where plants and animals live. The set of physical, chemical, biological, and geologic conditions under which the original sediments of a given rock layer were laid down are called the *depositional environment.*

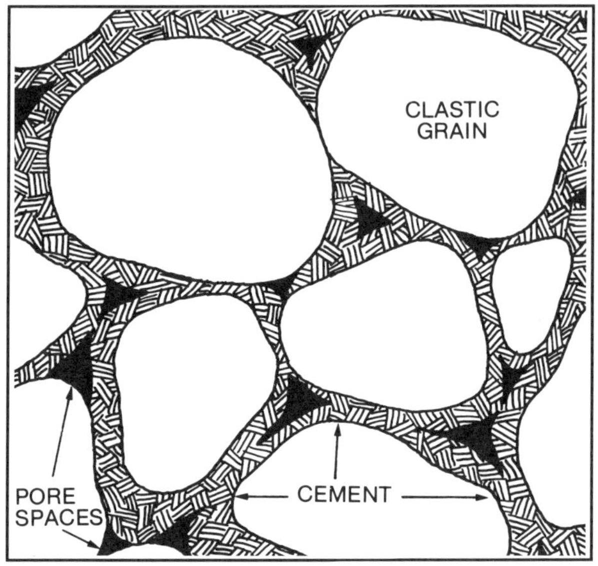

Figure 10. Cementation of sediment

Depositional Environments

Sediments accumulate in characteristic patterns and locations relative to the continental masses. As a continent moves away from a midocean ridge, its trailing edge subsides; here, thick layers of clay from the land and lime mud from marine organisms accumulate in the shallow sea on the *continental shelf* (fig. 11*A*). The advancing edge of a continent may be crumpled and broken in mountainous *fold belts* and *overthrust belts;* coarse, jumbled gravels from these mountain ranges accumulate in the adjacent lowlands (fig. 11*B*). In places, the crust is pulled apart by deep-rooted forces and forms down-dropped basins called *grabens* (fig. 11*C*); here,

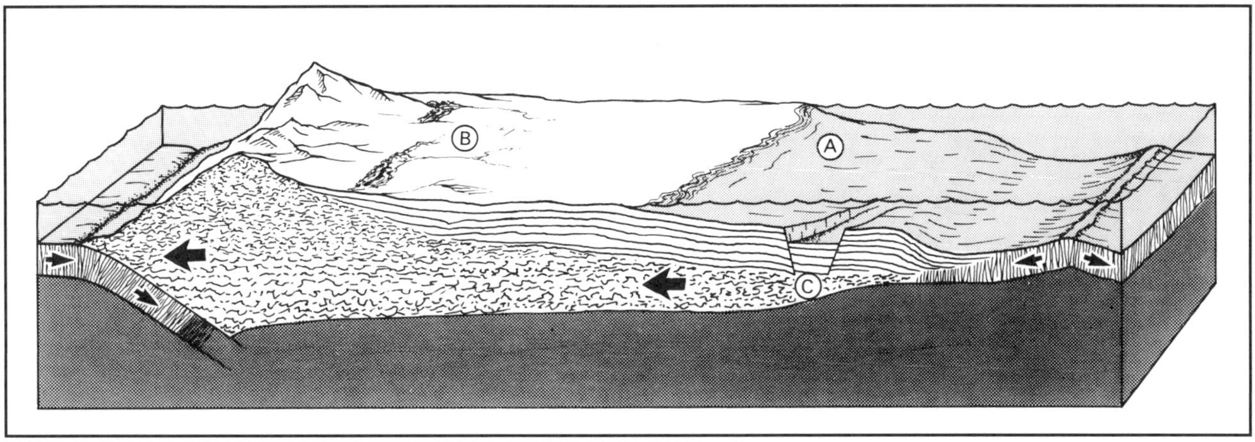

Figure 11. Depositional environments: *A*, continental shelf; *B*, continental lowlands; *C*, graben on continental shelf

sediments may accumulate to depths of several miles as the basin deepens.

In any sedimentary basin, the type of sediment that accumulates depends largely on the energy of the water that deposits it: higher energy means larger grains. A fast-flowing, energetic stream carries off small particles, leaving behind coarser sediments such as gravel and boulders. Thus, coarse sediment indicates a high-energy depositional environment. The variability of the energy level affects the uniformity of grain size—that is, *sorting*. An unseparated collection of different-sized particles is said to be unsorted or poorly sorted. A dry desert arroyo where flash flooding sometimes occurs tends to collect a jumble of unsorted sediments (fig. 12); a steady stream that flows year-round deposits well-sorted layers of particles similar in size and shape (fig. 13). *Grading* is an indication of a variable energy level, as in a wet/dry climate cycle. A stream may flow swiftly in the wet season, depositing coarse sediments, then gradually slacken, overlaying a succession of ever-finer materials (fig. 14).

A classification scheme for depositional environments is shown in table 1. Each environment listed can include many types of sediments, but each environment has a characteristic assemblage of types. Typical stream deposits, for example, range from gravel and boulders in areas of high flow velocity and turbulence to fine silt and clay in the floodplain flanking the main channel. Deposition along a streambank is characterized by an overlapping series of sandbars and clay sheets, with ripple marks and

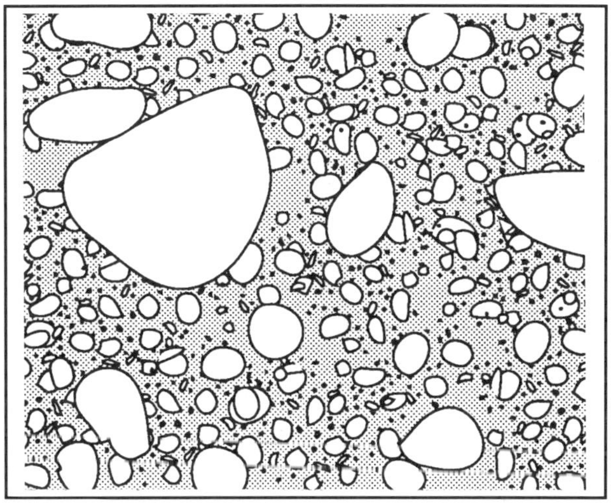

Figure 12. Unsorted sediments

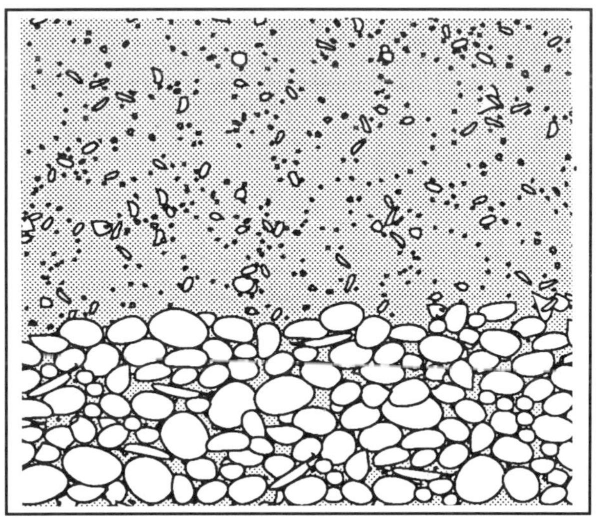

Figure 13. Sorted sediments

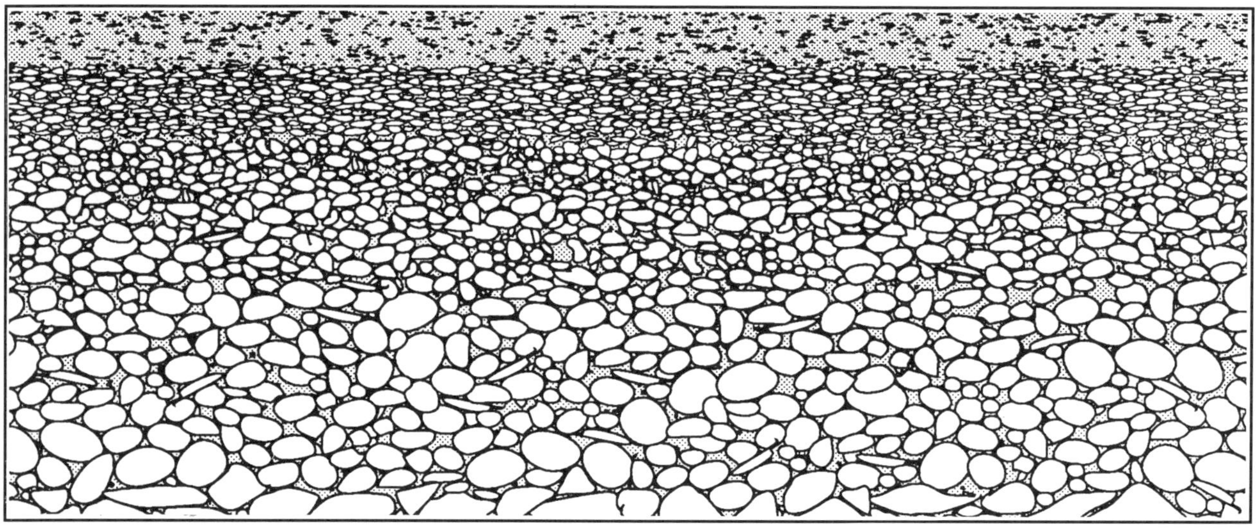

Figure 14. Graded sediments

other flow features preserved on the top of each layer (fig. 15). An intermittent desert or mountain stream that is prone to flash flooding typically dumps its load of unsorted materials in an *alluvial fan,* a chaotic jumble of boulders, gravel, sand, and clay found where the gradient flattens out (fig. 16).

The energy level along a beach is moderate and relatively constant. Wave action suspends fine clay particles, carrying them out to quieter offshore areas, but leaves clean, well-sorted sand at the water line (fig. 17). High-energy (coarse) deposits are concentrated in the surf zone and in the backshore zone between high tide and storm tide levels; low-energy, fine sediments occur seaward of the shoreface and in sheltered lagoons.

The depositional environments shown in table 1 grade into one another in a variety of ways. For example, the wind is a significant depositional factor not only in deserts far from the sea, but also along many of the world's seacoasts, where it piles sand into great dunes beyond the reach of the tide. A

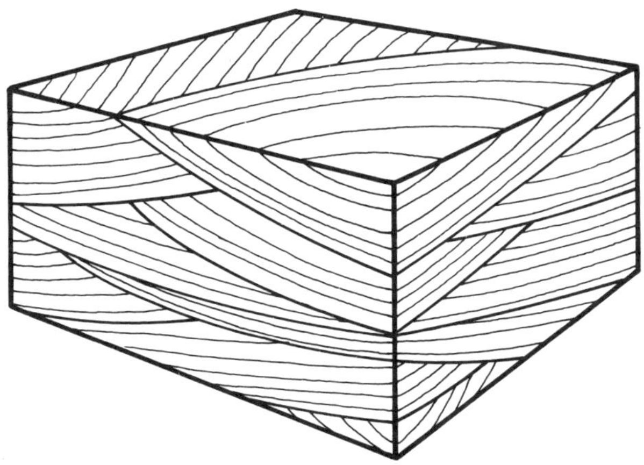

Figure 15. Depositional layers along a stream bank

Table 1
Depositional Environments

Continental	Transitional	Marine
Fluvial (stream)	Delta (river mouth)	Reef
Lacustrine (lake)	Interdeltaic shoreline (beach)	Continental shelf
Aeolian (wind)		Continental slope
Glacial (ice)		Continental rise

Figure 16. Alluvial fan

sabkha (also called a playa or a sebkha) is a shallow desert basin where infrequent runoff collects and evaporates, leaving thin alternating layers of clay and evaporites (salt, gypsum, and other soluble minerals); sabkhas are found both far inland and along arid coastlines. Fluvial and beach sediments often overlap and interweave as shifting shorelines are cut by rivers; both are characterized by sandbars, but the orientation of these deposits and the shape and arrangement of their sand grains differ.

When a river reaches the coastline, its flow energy dissipates in the sea. No longer able to transport its solid load, the stream deposits sediments in a *delta.* A typical *marine delta* is a fan-shaped body of sediments projecting beyond the normal coastline (fig. 18). Its *top-set* beds, essentially a seaward extension of the stream

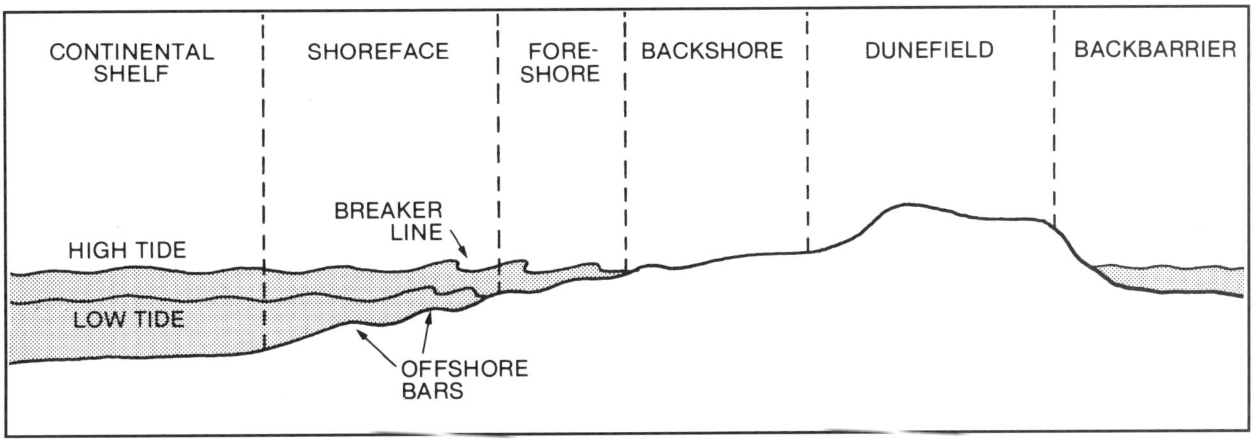

Figure 17. Depositional environments of the seashore

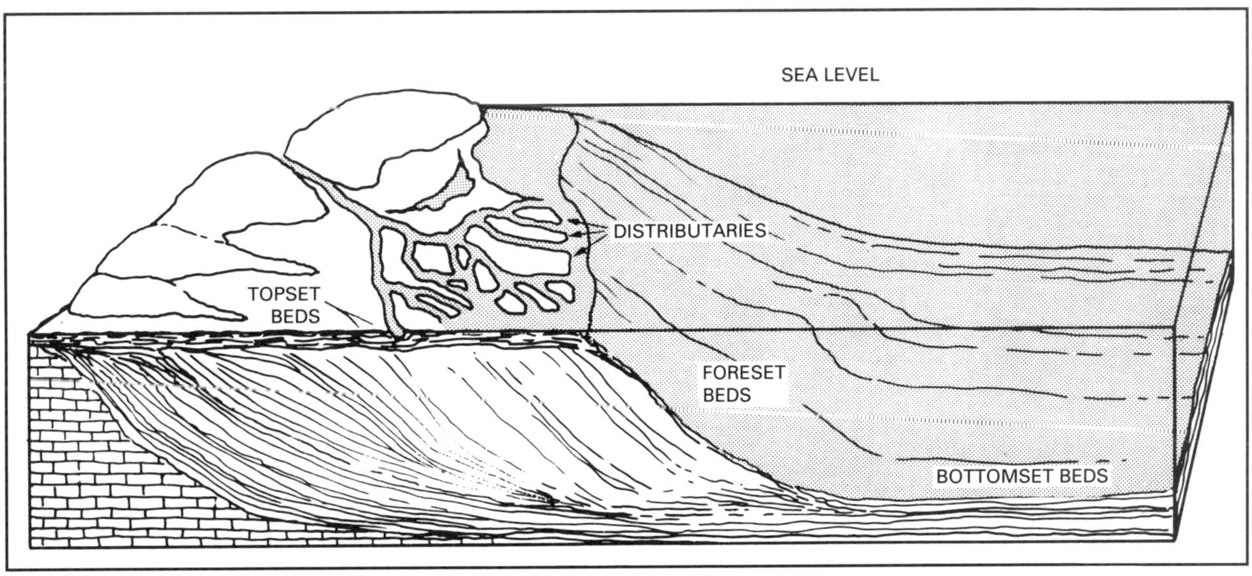

Figure 18. Marine delta

channel, contain clay, silt, sand, and gravel in patterns much like those of the continental lowland drainage. *Foreset* beds, on the steep seaward face of the delta, are composed of silt and clay; *bottomset* beds, beyond the delta face where the river's flow energy is finally dissipated in the sea, consist mostly of fine clays. The same general pattern prevails in a *lacustrine* delta, where a stream enters a freshwater lake.

On a growing *(prograding)* delta, the river's mouth may shift from one part of the delta to another as sediments block channels and flow seeks the easiest path across it. This shifting causes a delta to build up in a series of overlapping lobes (fig. 19). In the last 5,000 years, the active lobe of the Mississippi Delta has shifted from what is now the mouth of the Atchafalaya River to its present location southeast of New Orleans.

The continental shelf, a true marine environment beyond the deltas and beaches of the transition zone, is characterized by fine clastic sediments, often with an abundance of organic material, that form shale (fig. 20). In warmer climates, carbonate muds may accumulate to great thicknesses to form limestone.

Figure 20. Shale

Such marine rock assemblages now account for most of the world's known petroleum reserves. One of the largest oilfields, the Ghawar in Saudi Arabia, occurs in a folded limestone formation.

At times in the geologic past, shallow arms of the sea extended far into the interior of the continents

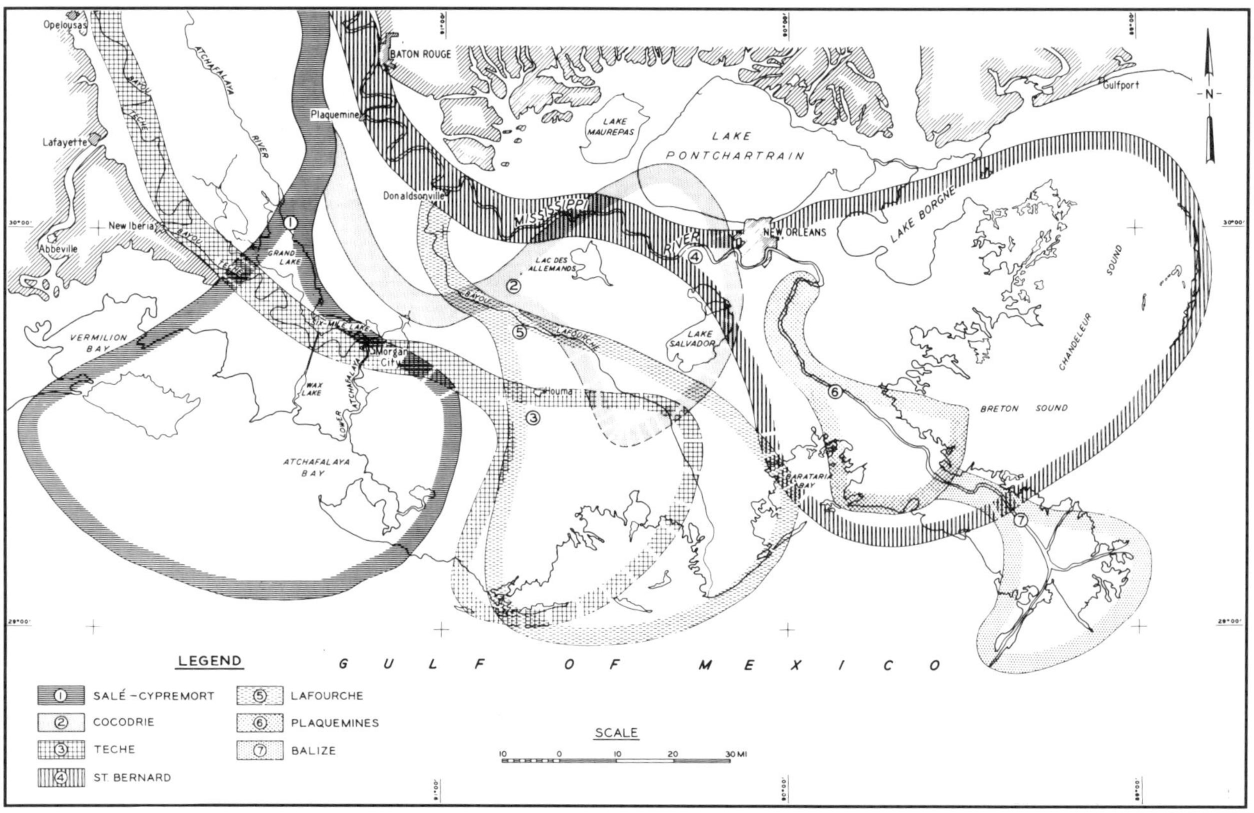

Figure 19. Mississippi River delta (after C. R. Kolb and J. R. van Lopik)

(fig. 21). Sedimentation patterns in these *epeiric seas* were much like those of the present continental shelves. Some of the inundated areas, however, were so far from the open sea that they were almost landlocked, like the present Baltic Sea of northern Europe. Epeiric seas near the equator, warmed by the tropic sun, supported rich communities of marine plants and animals, which later contributed their organic material to the formation of thick sequences of shales and limestones. As water evaporated and was replaced by inflow from the sea, salt concentrations often rose so high that salt precipitated out to form salt beds on the seafloor.

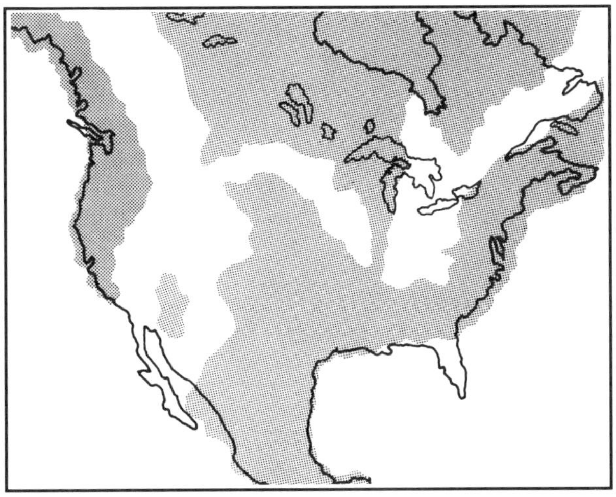

Figure 21. Epeiric seas of North America in the Paleozoic era

Types of Sedimentary Rock

Each depositional environment has its characteristic assemblage of sedimentary rock types. When discussing these types, it is convenient to think in terms of three basic types: clastics, carbonates, and evaporites. Note, however, that any rock is likely to have characteristics of more than one of these types.

Clastics

Clastic sedimentary rocks are composed mostly of particles derived from other rocks. There are two basic types of clastic particles: *mineral* grains, composed entirely of a single mineral, such as quartz, feldspar, or mica; and *lithic* grains, which consist of an assemblage of different minerals, like miniature rocks. In rocks with *clastic texture,* the grains touch each other but do not interlock. The *crystalline texture* of igneous rock, by contrast, is characterized by mineral grains that are in contact on all surfaces, having formed and grown together as the rock solidified. Sedimentary rock usually has empty (or fluid-filled) spaces between grains (fig. 22).

Clastic rocks are classified primarily by grain size (table 2). They are named according to the size of the particles that make up more than 50 percent of their bulk. A rock composed of 60 percent sand and 40 percent calcite, for example, would be called a limy sandstone.

The coarsest rocks, *conglomerates,* indicate former high-energy environments: steep topography, swift streams, heavy surf (fig. 23). Some conglomerates are made up of broken, angular particles that have not been rounded and smoothed by transport. These

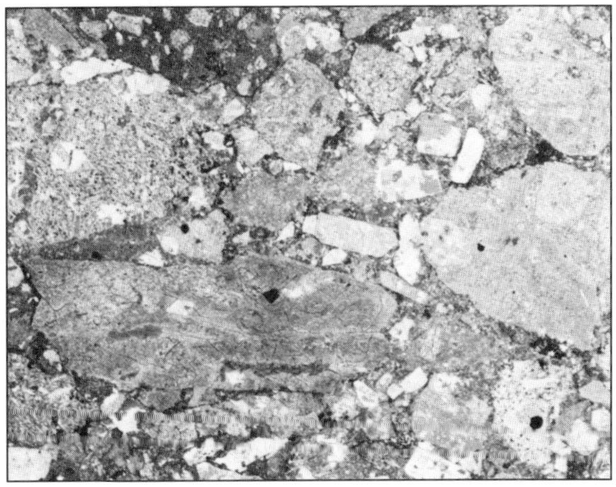

A

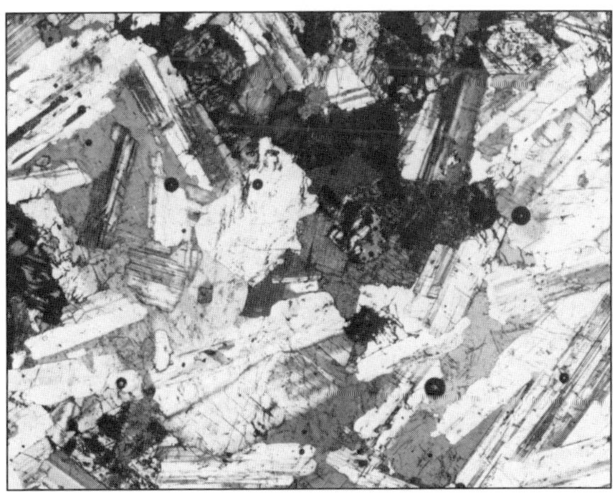

B

Figure 22. Clastic (*A*) and crystalline (*B*) texture

Table 2
Clastic Sedimentary Rocks Classified by Grain Size

Particle Name	Diameter Range	Rock Type
Gravel	Larger than 2 mm	Conglomerate
Sand	$\frac{1}{16}$ mm–2 mm	Sandstone
Silt	$\frac{1}{256}$ mm–$\frac{1}{16}$ mm	Siltstone
Clay	Smaller than $\frac{1}{256}$ mm	Shale

Conglomerate

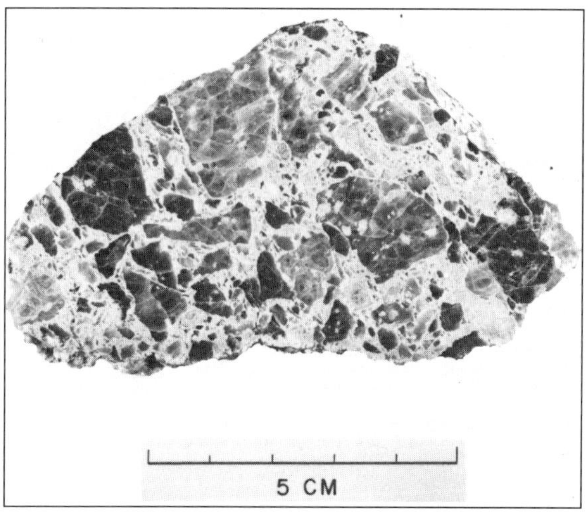

Breccia

Figure 23. Conglomerates

rocks, called *breccia,* are typical of landslides, volcanic debris, and certain glacial deposits.

About 25% of the world's sedimentary rock is sandstone, composed mostly of particles $\frac{1}{16}$–2 millimeters in diameter. Sandstones vary widely in mineral content, grain shape, sorting, and other characteristics; the cleanest, most uniform, most porous sandstones are those deposited in beach and dune environments. Well-sorted sandstones with round, smooth grains tend to be very porous; about one-third of their bulk may be void space (fig. 24). Porosity can be reduced by the infiltration of finer sediments, by cementation, and to a limited extent, by compaction. It can be increased by the leaching out of cement or individual mineral grains, or by the removal of fine particles by groundwater. The texture of *siltstone* is similar to that of sandstone, but the grains of the finest siltstones are too small to be seen by the unaided eye.

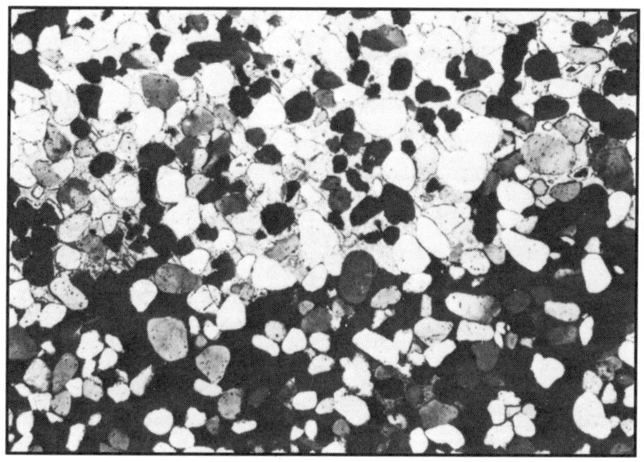

Figure 24. Sandstone

Unlike the generally round grains of sandstone, the flat, microscopic particles of clay that make up a typical *shale* are both adhesive and cohesive; that is, they cling to one another and to water, making clay both sticky and water-absorbent. The clay particles in a freshly deposited layer have a loose, disorderly arrangement, like a heap of cards (fig. 25*A*). Such a deposit may have a porosity of 90% or more and contain a great deal of water. When deeply buried and compacted, however, clay particles break and line up like bricks in a wall with little void space between (fig. 25*B*). Porosity may be reduced to 10% or less as fluids are squeezed out.

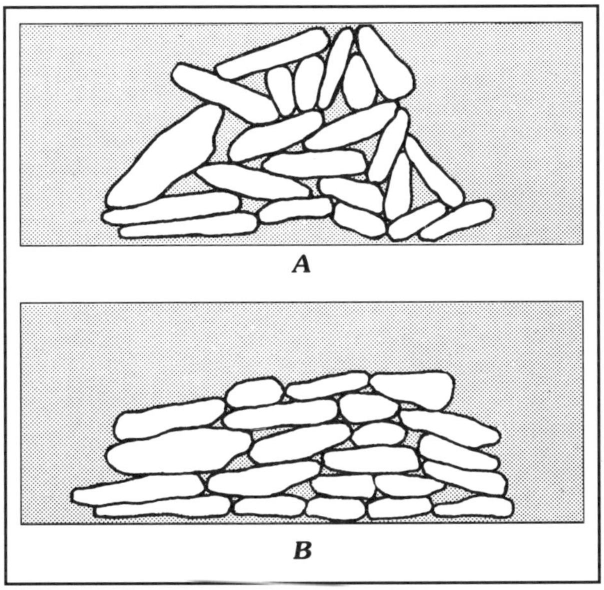

Figure 25. Freshly deposited (*A*) and deeply buried and compacted (*B*) clay

Carbonates

The carbonates, sedimentary rocks that consist mostly of calcium carbonate and magnesium carbonate, are *limestone* and *dolomitic limestone* (often called simply *dolomite*). They are formed by any of several processes or a combination. One of the most important of these is a life process; for this reason, limestone is sometimes classified as an *organic* rock.

Many marine organisms take calcium from the water and use it to make a shell. When these organisms die, their shells fall to the bottom and accumulate along with mineral grains, typically the clay that is deposited in quiet backwaters, where life is most abundant. The result is *lime mud*, a calcite-rich sediment that is the starting point for shaly limestone. Limestone often contains an abundance of fossils, especially the shells of calcareous organisms.

Oolitic limestone is composed largely of the rounded sandlike grains of calcite known as *ooliths,* formed by the accretion of layers of calcium carbonate on smaller particles, like scale in a boiler. A cross section of an oolith reveals an internal structure much like that of a hailstone (fig. 26).

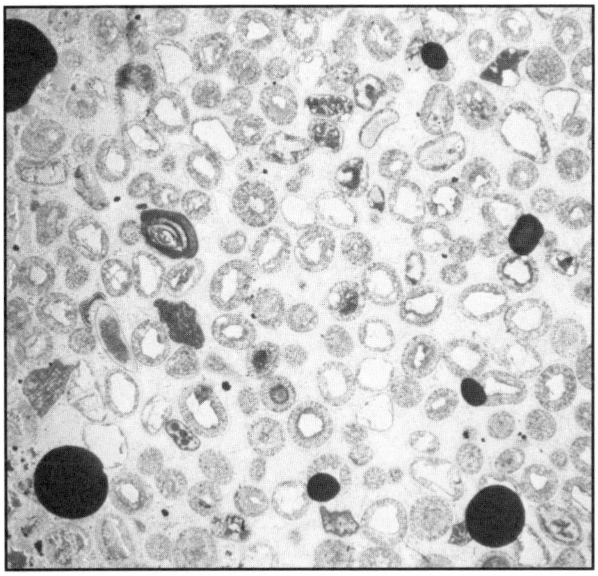

Figure 26. Oolitic limestone

Reef limestone is formed more or less in place from the skeletons or shells of large colonies of marine animals (fig. 27). A coral reef, for instance, is made up of the branching skeletal remains of large colonies of tropical coral polyps, on which other skeletal debris and shell fragments have accumulated. El Capitan peak in West Texas is a Permian reef that was buried in an epeiric sea and later uplifted and exposed (fig. 28).

The porosity of limestone is little affected by compaction, but depends largely on the type and proportion of other sediments, such as clay or sand, that make up the rock, as well as on the degree to which calcite or other cement fills its pores. New limestone is very porous—as much as 60% to 70% void space. As limestone ages, cementation can reduce porosity to 5 percent or less. Later leaching by aerated groundwater may restore lost porosity by creating solution channels and small caverns called vugs. Leaching by magnesium-rich water can also

Figure 27. Model of Permian Basin coral reef

Figure 28. El Capitan Peak in West Texas

lead to *dolomitization,* the replacement of calcium carbonate by magnesium carbonate (dolomite). The porosity of limestone petroleum reservoirs ranges from 5% to 20%, but is usually localized and irregular.

Evaporites

A third type of sedimentary rock is formed from the dissolved minerals left behind when water evaporates. Halite, rock salt, is one of the most common evaporites. Deep beneath the seafloor in the Gulf of Mexico lie thick beds of salt that were deposited millions of years ago when seawater evaporated from an isolated ocean basin. As the basin deepened, clastic sediments were laid down over the salt. The weight of these overlying sediments has deformed the soft, light salt layer, causing it to bulge toward the surface in a series of mushroomlike columns (fig. 29). Although the salt is nonporous and thus cannot contain oil or gas, each column pushes overlying porous layers upward in petroleum-trapping domes.

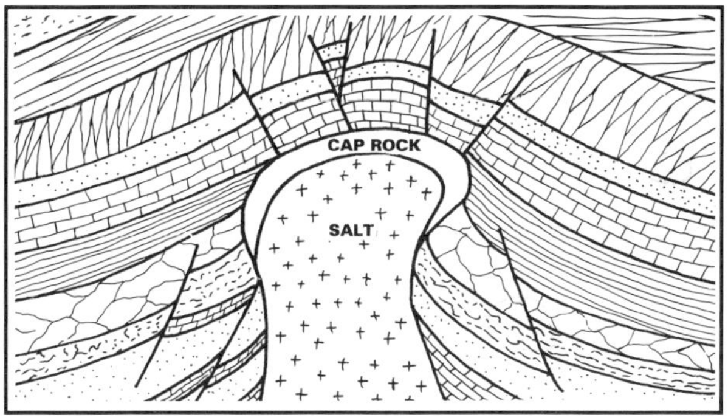

Figure 29. Salt dome

REVIEW QUESTIONS

Chapter 2

_____ 1. All the processes involved in transforming sediments into rock are included in the term—

- A. cementation.
- B. diagenesis.
- C. lithification.
- D. compaction.

2. The set of physical, chemical, biological, and geologic conditions under which sediments are deposited is called the _____ _____.

_____ 3. Sandstone accounts for about _____ % of all sedimentary rock.

- A. 5
- B. 10
- C. 25
- D. 50

_____ 4. Which of the following is an evaporite?

- A. halite
- B. granite
- C. shale
- D. trilobite

_____ 5. Vugs are—

- A. volcanic lava formations.
- B. small cavities in limestone.
- C. intergranular voids in sandstone.
- D. faulted sedimentary basins.

_____ 6. Which of the following rock types requires the least depositional energy?

- A. shale
- B. conglomerate
- C. sandstone
- D. breccia

_____ 7. A layer of gypsum is most likely to have been deposited in which of the following climates?

- A. warm and rainy
- B. cold and dry
- C. cold and snowy
- D. hot and dry

_____ 8. If 65% of the volume of a rock consists of particles smaller than $\frac{1}{256}$ mm and 35% of particles $\frac{1}{256}$–$\frac{1}{16}$ mm, the rock would be called—

- A. shaly siltstone.
- B. limy sandstone.
- C. silty shale.
- D. sandy siltstone.

_____ 9. Rounded, multilayered calcite grains are—

A. ooliths.
B. fossils.
C. vugs.
D. reefs.

_____ 10. (T/F) A well-sorted sandstone is likely to be more porous than a poorly sorted one.

_____ 11. Dolomite is a carbonate rock in which ____________ has replaced more than half of the calcium.

A. sodium
B. salt
C. clay
D. magnesium

_____ 12. A playa deposit consisting of alternating thin layers of clay and evaporite is a ____________.

_____ 13. Lithic clastic particles are composed of—

A. quartz.
B. sand.
C. feldspar.
D. an assemblage of minerals.

_____ 14. Which of the following types of rock is most affected by compaction?

A. sandstone
B. limestone
C. shale
D. halite

_____ 15. (T/F) Rock salt is denser than most sedimentary rocks.

CHAPTER 3
STRATIGRAPHY

Introduction

As geologists gather information about the structure and character of the rock formations, they arrange it in graphic form. This not only helps them visualize what they cannot see directly, but also provides a way to communicate their findings to others. In particular, it helps them visualize and display the *stratigraphy* of the crust—information that describes the origin, composition, distribution, and succession of rock layers.

Maps, Sections, and Diagrams

Geologic information can be arranged graphically to show variation either horizontally or vertically. Horizontally arrayed data are *maps;* vertically arrayed data are *sections.* Sometimes these two types of arrays are combined in a simulated three-dimensional graph.

A geologist usually starts with a *base map* of the area of interest, showing survey benchmarks, property lines, and such nongeologic surface features as streams, roads, and buildings. Base maps are useful for planning exploration, leasing, road building, well placement, and other activities.

Another type of surface map, a *topographic map,* displays land elevations as a series of contour lines, each line connecting points of equal elevation (fig. 30). A topographic map shows the geologist the shapes of hills and valleys, often revealing subsurface geologic structure by the pattern of exposed and eroded formations. The surface is described in a different way by an *outcrop map,* which shows the rock types at the surface or just beneath the layer of soil (fig. 31). Elevations and rock types are sometimes shown on the same map.

A contour map showing elevations of a subsurface rock layer or structure is called a *structure contour map* (fig. 32). This type of map is particularly useful in petroleum exploration because it shows, among other things, the topographic highs in porous formations where oil and gas are most likely to accumulate. Similarly, formation thick-

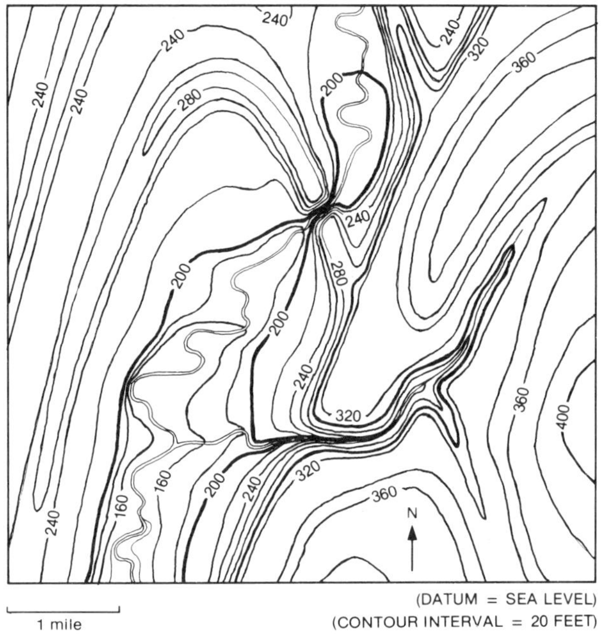

Figure 30. Topographic contour map

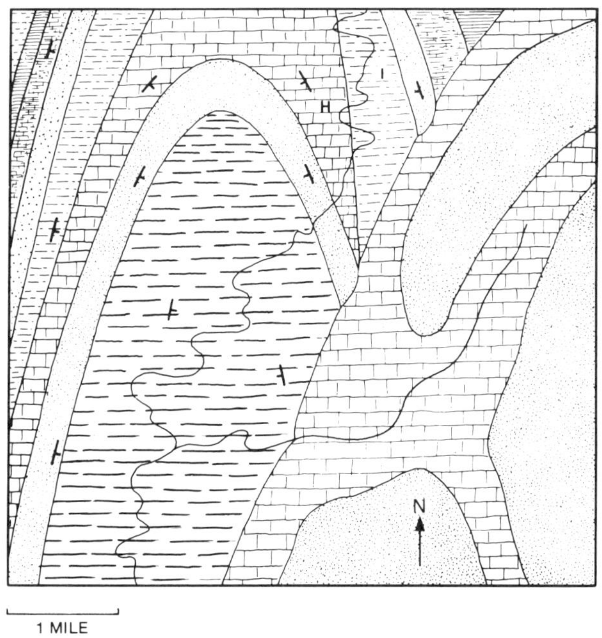

Figure 31. Geologic outcrop map

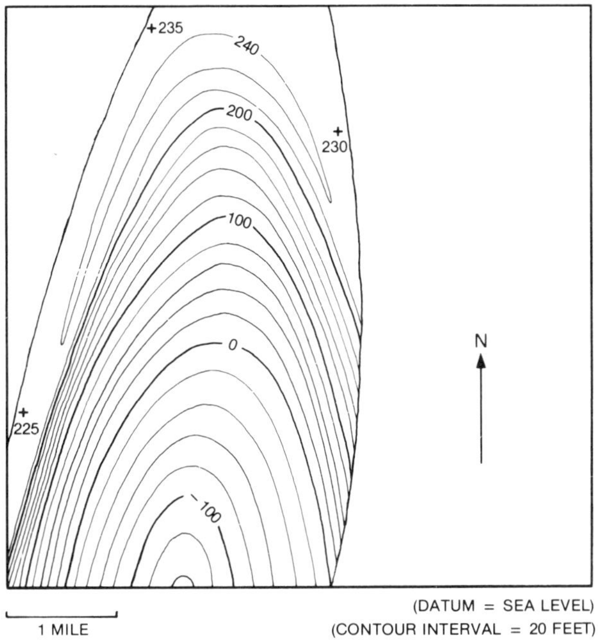

Figure 32. Structure contour map

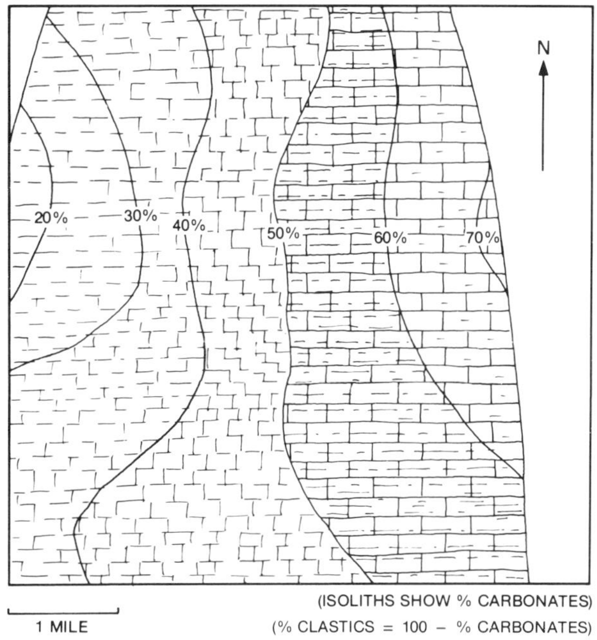

Figure 34. Lithofacies map

nesses can be delineated by an *isopach map,* on which contour lines connect points of equal formation thickness (fig. 33).

Formation characteristics such as porosity, permeability, grain size, clay content, and cementation change from one location to another. A map showing such variations is called a *lithofacies map.* Figure 34 is a lithofacies map of a reservoir formation in which carbonates range from less than 20% to more than 70% of total rock volume. A petroleum geologist would find this map helpful in locating the part of the reservoir with the greatest production potential.

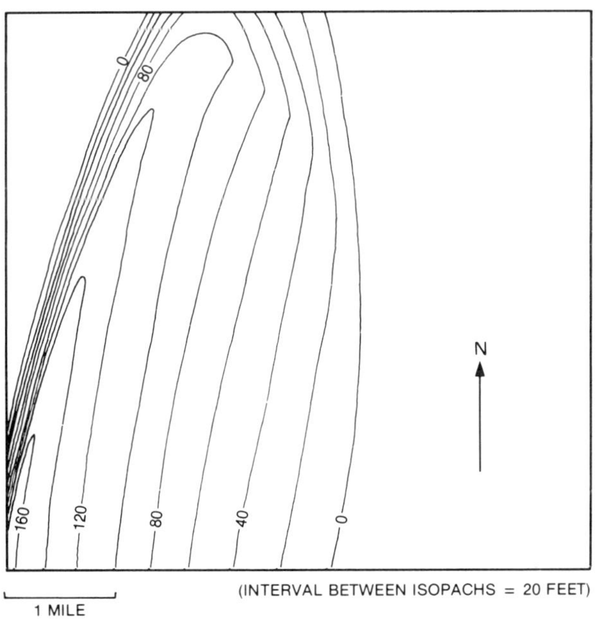

Figure 33. Isopach map

A *section* is a cutaway view showing the sequence of rock layers beneath the surface (fig. 35). Different formation characteristics can be displayed in a section, including rock type, vertical thickness, and elevation. A section can be constructed along either a straight line or a line connecting a series of points not in a straight line, such as drilled wells.

A *block diagram* combines the vertical presentation of one or more sections with the horizontal data normally displayed on a map. It may look very realistic, as though a piece of the crust were lifted out for examination (fig. 36). Usually, though, it is more abstract. A block diagram is useful for showing three-dimensional geologic data in the two dimensions of a sheet of paper or a video display terminal.

Principles of Stratigraphy

As mentioned before, *stratigraphy* is the study of the origin, composition, distribution, and succession of rock layers. The term encompasses the techniques used by the geologist to determine the succession of depositional environments and the relative ages of rocks.

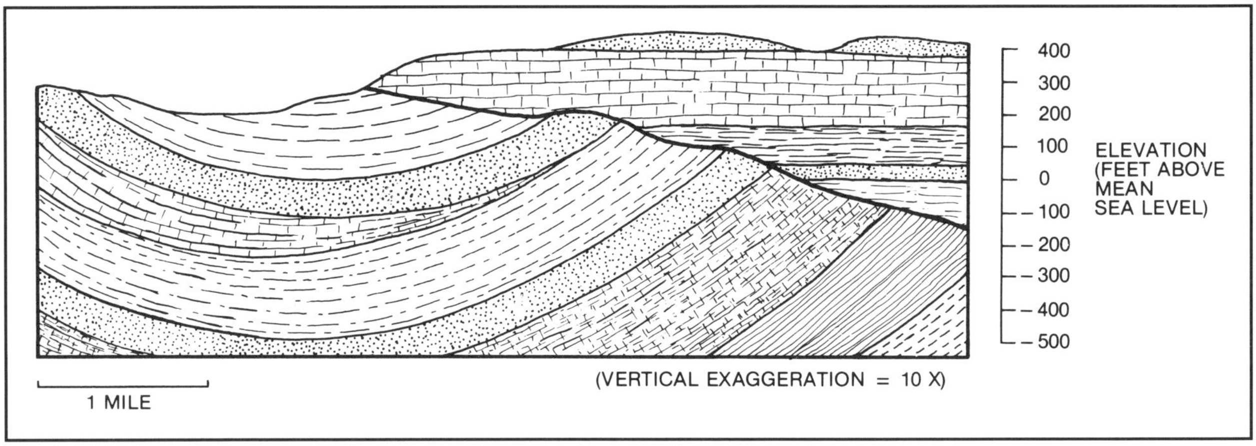

Figure 35. Section

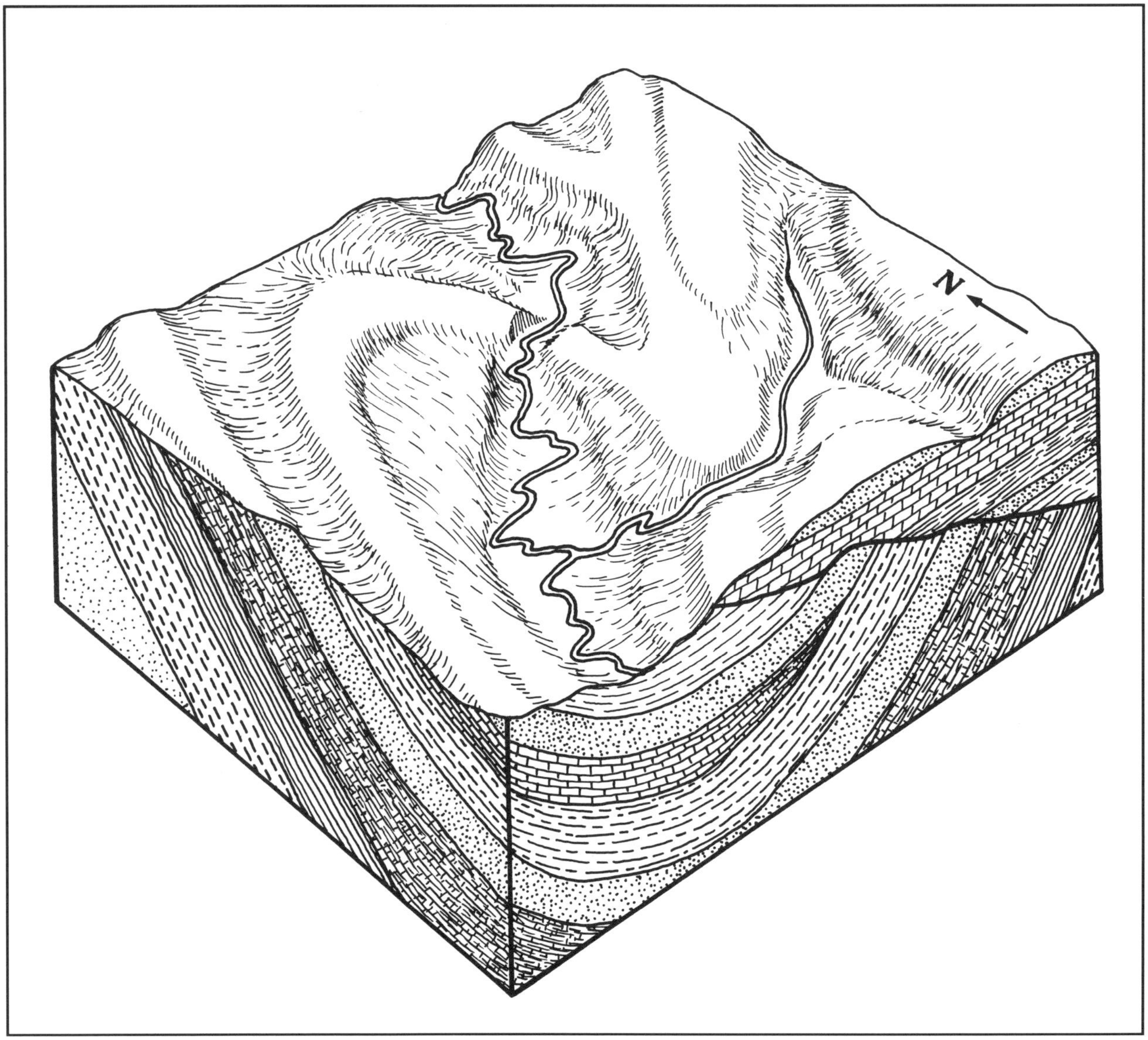

Figure 36. Block diagram

A sedimentary layer is deposited in a continuous, unbroken sheet with an essentially horizontal upper surface but a lower surface that conforms to a previous land surface or seafloor (fig. 37). Its edges "lap out" like water at the shore of a lake. Each layer is deposited on top of older sediments; in an undisturbed series of rock layers, the youngest layer is at the top and the oldest is on the bottom.

Once a sedimentary rock layer is formed, its continuity or horizontality may be disrupted in a number of ways: it may be partly eroded; it may be fractured, faulted, bent, or folded by crustal movement; it may even be invaded by magma that cools to form a subsurface body of igneous rock. Just as any formation is known to be younger than the formation beneath, any event that disrupts the continuity or horizontality of a formation is more recent than the deposition of the layer affected. Similarly, any layer that cuts across another layer must have been deposited after the layer it cuts and is therefore younger (fig. 38).

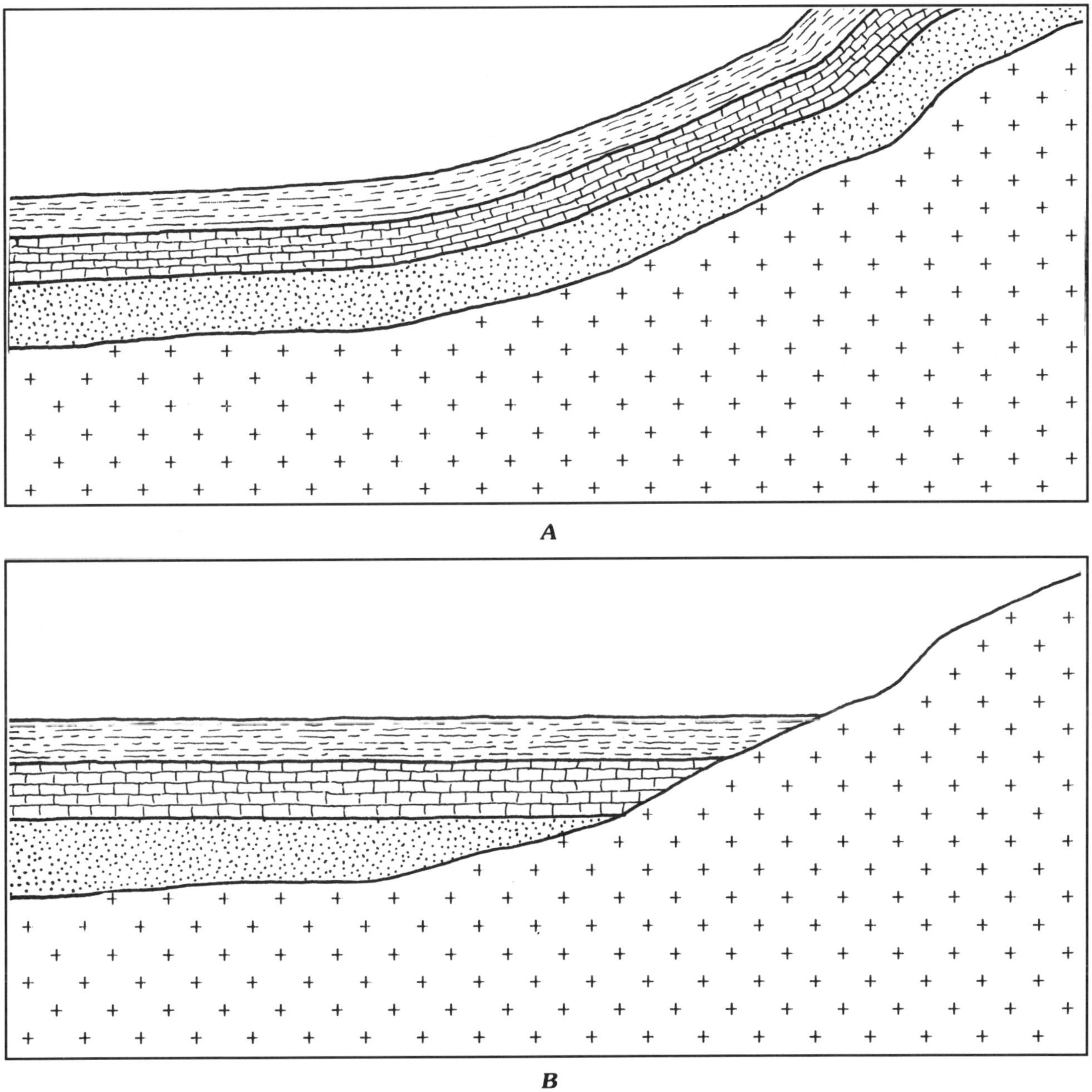

Figure 37. Law of original horizontality: sedimentary layers are formed not as in *A*, but as in *B*

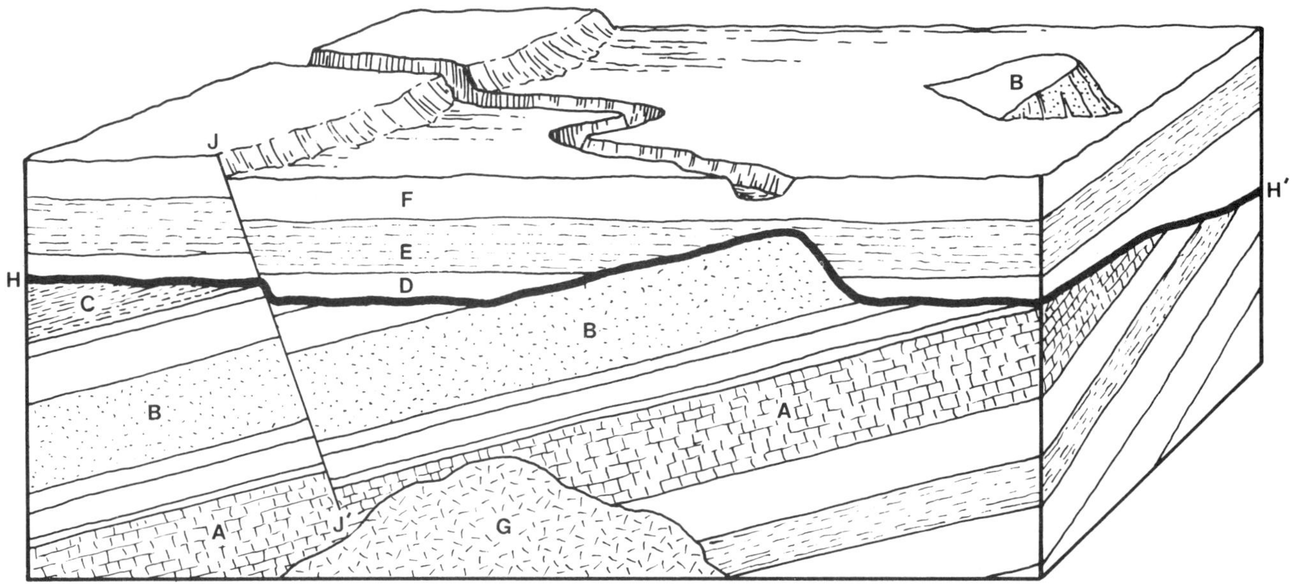

Figure 38. Block diagram showing relative ages of features: basement rock G is older than layers A, B, and C, which are older than D, E, and F; angular unconformity H-H′ is older than fault J-J′

Folding and Faulting

Tectonic plate motion is one of the events that can change the shape and orientation of sedimentary rock layers. Wherever plates converge, the crust is subjected to enormous horizontal forces that can gradually compress it by dozens or even hundreds of miles, wrinkling and folding it like a giant throw rug (fig. 39). Each upfold of the crust is an *anticline;* each downfold is a *syncline* (fig. 40).

Figure 39. Rock layers folded by horizontal compression

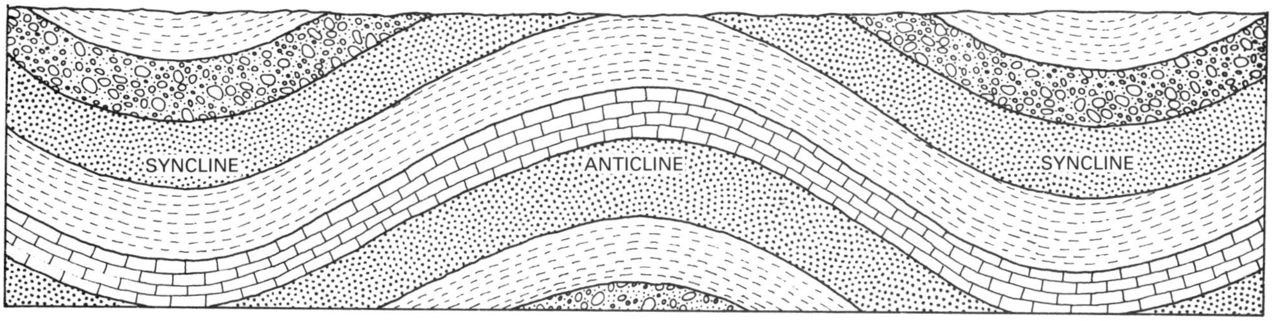

Figure 40. Anticline and synclines

Anticlines and synclines are graphic proof that solid rock can flow like the ice in a glacier. Like most solid materials, rock is slightly plastic; under uniform pressure over long periods, it will bend without breaking. However, if stress is applied unevenly or if it exceeds the rock's breaking strength, the rock fractures. A fracture in the crust along which the rocks on opposite sides have shifted relative to each other is termed a *fault.*

A *normal fault* is one whose slip plane is at a steep angle with the surface and along which the rock on the upper side has slipped downward in the direction of the dip (fig. 41*A*). (In geology, the *dip* of any surface is the direction in which a marble would roll if placed on it.) A *reverse, or thrust, fault* (fig. 41*B*) is one in which the rock on the upper side has been displaced upward along the fault plane. A normal fault allows extension of the crust; it is often caused by forces that stretch the crust. A thrust fault is caused by forces that squeeze the crust together, causing a break where one piece overrides another.

An *overthrust fault* is a thrust fault whose slip plane is nearly horizontal; its displacement is the result of large horizontal movements of the crust (fig. 42). Along some overthrust faults, one slab has slipped several miles over the top of another so that a well drilled through the fault would penetrate the same series of rock layers twice.

The opposite sides of a *lateral,* or *strike-slip, fault* move horizontally past each other; the fault plane itself may be vertical (fig. 43). The most familiar example is California's San Andreas fault, where the Pacific plate is slipping northward about 2 inches per year past the edge of the North American plate. If this

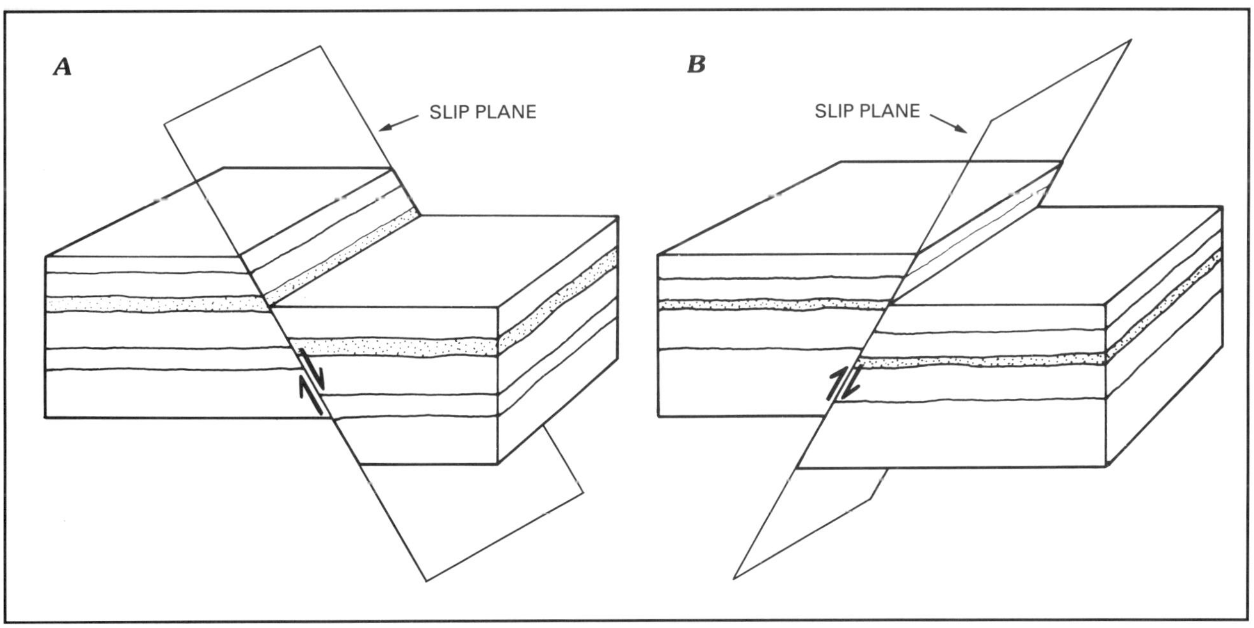

Figure 41. Normal (*A*) and reverse (*B*), or thrust, faults

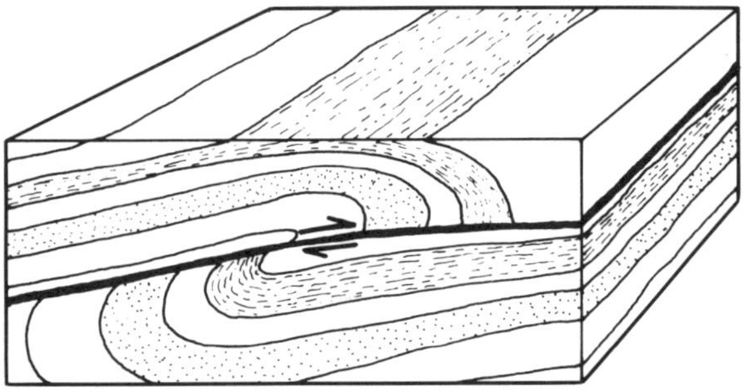

Figure 42. Overthrust fault

slip occurred continuously, the San Andreas fault would be merely a geologic curiosity. Instead, sections of the fault "lock up" for years, releasing the strain all at once in a sudden, powerful earthquake.

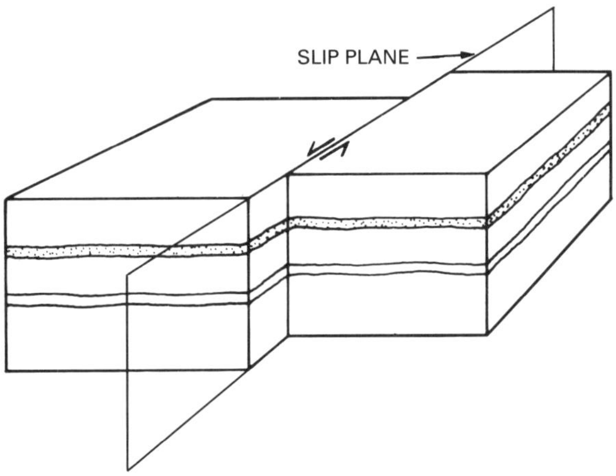

Figure 43. Lateral, or strike-slip, fault

One type of fault common along the Gulf Coast is the *growth*, or *rollover, fault* (fig. 44). Often invisible at the surface, a growth fault is an active slip plane in unconsolidated sediments where continued deposition causes layers on the downthrown side to grow thicker than those across the fault. The plane of a growth fault curves toward the horizontal at depth, and total displacement at depth is greater than near the surface. Curvature of the layers on the downthrown side often creates a broad *rollover anticline.*

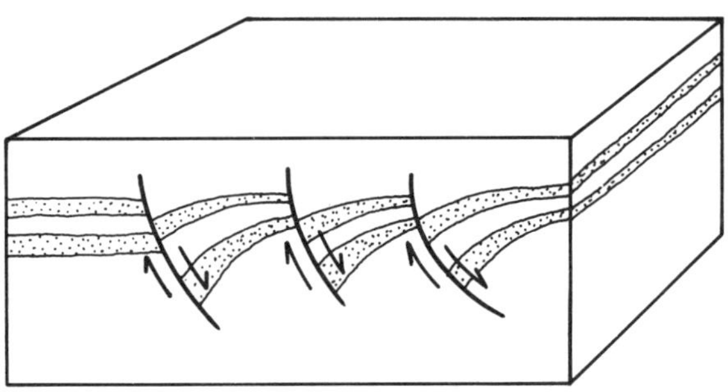

Figure 44. Growth, or rollover, faults

Unconformities

Sometimes a sedimentary basin is uplifted so that deposition ceases and erosion takes over. An *erosion surface* is formed; the upper surface of the most recent sediment layer, formerly smooth and horizontal, is modified by running water or other agents. Later, the region subsides and more sediments accumulate. If the succession of sedimentary layers is thought of as a chronological record, then the buried erosion surface represents a time gap of indeterminate length. Such a gap is called an *unconformity.*

There are several kinds of unconformity. If the uplift is gentle, so that the rock layers are not tilted or deformed, the gap in the geologic record is termed a *disconformity* (fig. 45). Although the layers of sediment above are parallel with those below, the shapes of ancient stream channels are often apparent in a disconformity (fig. 45*A*). However, the disconformity itself may be parallel with the layers above and below and therefore not readily apparent (fig. 45*B*).

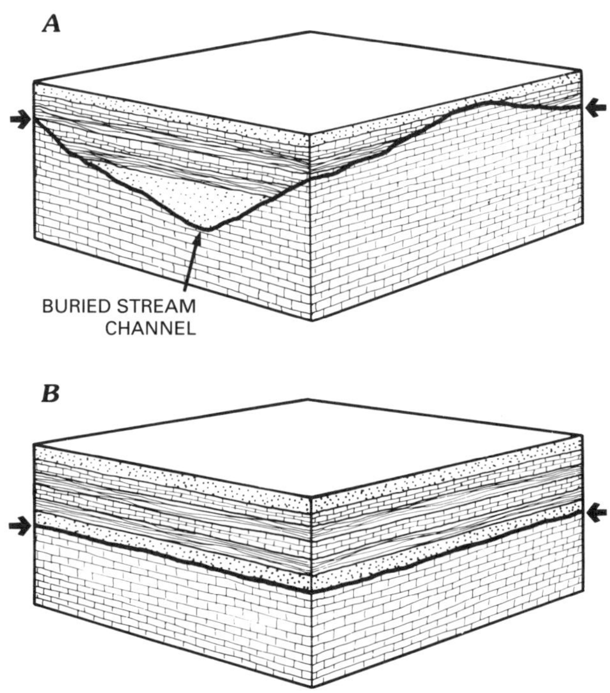

Figure 45. Disconformities

Deposition of sediments on layers that have been deformed and eroded produces an *angular unconformity* (fig. 46). Sedimentary layers below such an unconformity are not parallel with those above, and the gap in the record is obvious.

The most profound gap in the depositional record is the one beneath the oldest sedimentary layers. A *nonconformity* is an erosion surface on igneous or metamorphic rock that has been buried beneath sediments (fig. 47). In the geology of petroleum exploration, the rock beneath a nonconformity is usually referred to as *basement rock*.

Any unconformity is obviously younger than the rocks beneath it and older than those above. It is thus a useful tool in determining the relative ages of rocks and the events that have affected them over geologic time.

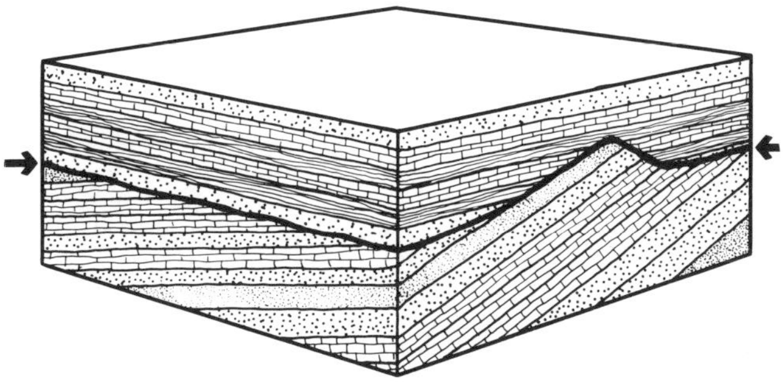

Figure 46. Angular unconformity

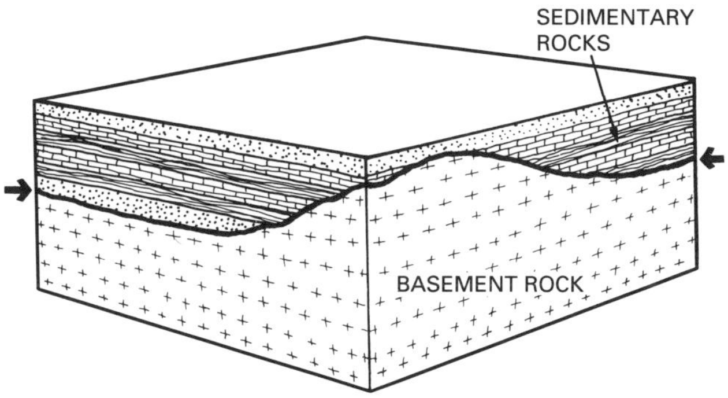

Figure 47. Nonconformity

REVIEW QUESTIONS
Chapter 3

____ 1. The study or description of the origin, composition, distribution, and succession of rock layers is called—

A. topography.
B. cartography.
C. stratigraphy.
D. photography.

____ 2. A ________________ shows three-dimensional geologic information in two dimensions.

A. block diagram
B. outcrop map
C. section
D. lithofacies map

____ 3. (T/F) The top of a new layer of sediment is horizontal or nearly so.

____ 4. In an overthrust fault, displacement is mostly—

A. horizontal.
B. vertical.
C. oblique.
D. strike-slip.

____ 5. (T/F) The rock beneath a nonconformity is always sedimentary rock.

6. A map showing nongeologic surface features that is used in planning oilfield exploration and development is a ____________ map.

____ 7. An isopach map shows the ___________ of rock layers.

A. porosity
B. clay content
C. outcrop
D. thickness

Use the diagram at right to answer the next three true/false (T/F) questions.

____ 8. Layer E is older than fault FF′.

____ 9. Layer B is older than layer D.

____ 10. Layer C is older than unconformity AA′.

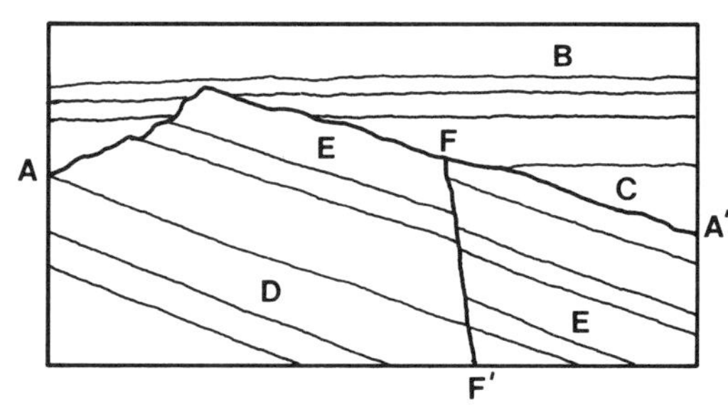

____ 11. An unconformity represents a surface on which—

A. faulting has been followed by erosion.
B. deposition has been followed by faulting.
C. subsidence has been followed by folding.
D. erosion has been followed by deposition.

B ____ 12. An upward fold or arch of sedimentary rock layers is called—

A. an overthrust fault.
B. an anticline.
C. a disconformity.
D. a syncline.

T ____ 13. (T/F) In a growth fault, the layers are thicker on the downthrown side.

T ____ 14. (T/F) An angular unconformity represents a greater time interval than a disconformity.

A ____ 15. An unconformity is always—

A. younger than the rocks beneath it and older than the rocks above it.
B. older than the rocks beneath it and as old as the rocks above it.
C. older than the rocks beneath it and younger than the rocks above it.
D. as old as the rocks beneath it and younger than the rocks above it.

CHAPTER 4
OIL AND GAS

Introduction

Geology concerns not only the form and structure of the earth, but also its past life forms and biotic communities as revealed in the fossil record. The evolution of living things, from the first simple organisms through the diverse and complex plants and animals of today, serves as a guide to the relative ages of the rocks in which their traces and remains are found.

In the search for petroleum, however, fossils are more than mere reference marks in the geologic time scale. They indicate what species and groups of organisms lived in a particular place at a particular time. Fossils and sediments also show whether the environment in which these associations of plants and animals lived was swamp, subtropical desert, or tropical ocean.

Why is the ecology of the past so important to the petroleum geologist? Because most of the evidence indicates that petroleum is derived from the remains of living organisms, especially small marine animals. Although there is a growing body of evidence that at least some gas came from deep within the earth's mantle, organic sources have been identified for most of the world's oil.

Petroleum is a mixture of chemical compounds called *hydrocarbons*—chains of carbon atoms with hydrogen atoms attached (fig. 48). These substances

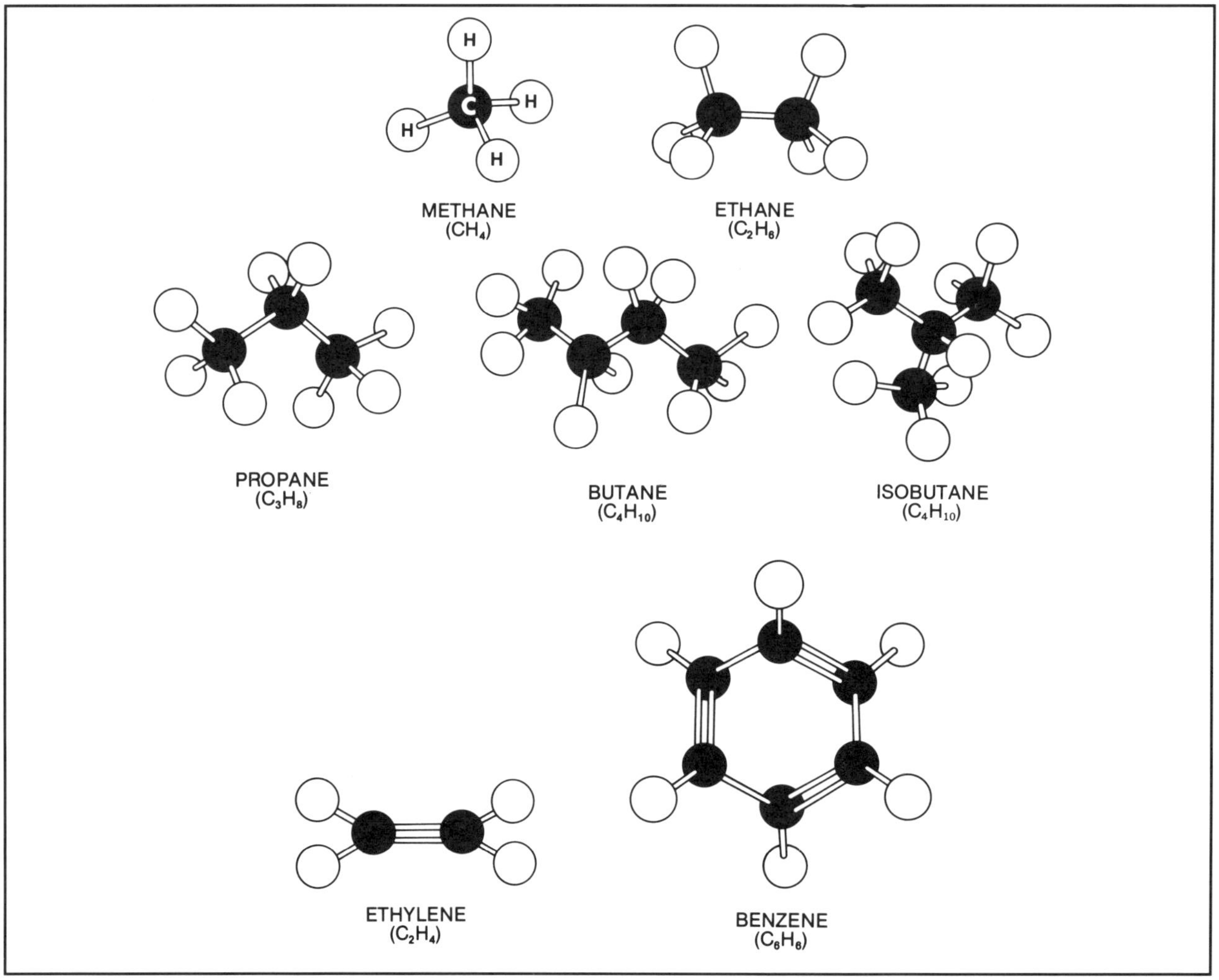

Figure 48. Hydrocarbon molecules

range from gas molecules, which have one to four carbon atoms, to liquids and solids with fifty or more carbon atoms. The simplest is the colorless, odorless gas *methane,* with one carbon atom and four hydrogens. The liquid known as *octane* has eight carbon atoms and eighteen hydrogens. Some molecules are branched or ring-shaped, such as *benzene,* a ring-shaped molecule with six carbon and six hydrogen atoms.

Source Beds

Most petroleum was formed in beds of shale that were deposited in shallow oceans with abundant marine life (fig. 49). Sheltered coastal waters, especially in tropical and subtropical regions, usually provide refuge for communities of life forms ranging from microscopic plants and animals, which constitute the bulk of the organic matter, to large predatory fish and mammals. Waste drifts down and accumulates on the seafloor along with dead organisms and uneaten remains. Rivers contribute organic detritus and sediments from the continent.

This organic matter consists mostly of carbon, hydrogen, and oxygen. As it drifts down, much of it is either eaten, and thereby recycled, or oxidized to form carbon dioxide and water. Whatever organic matter escapes being eaten or oxidized accumulates with clay and silt on the bottom. Here, anaerobic bacteria complete the decay process, breaking down

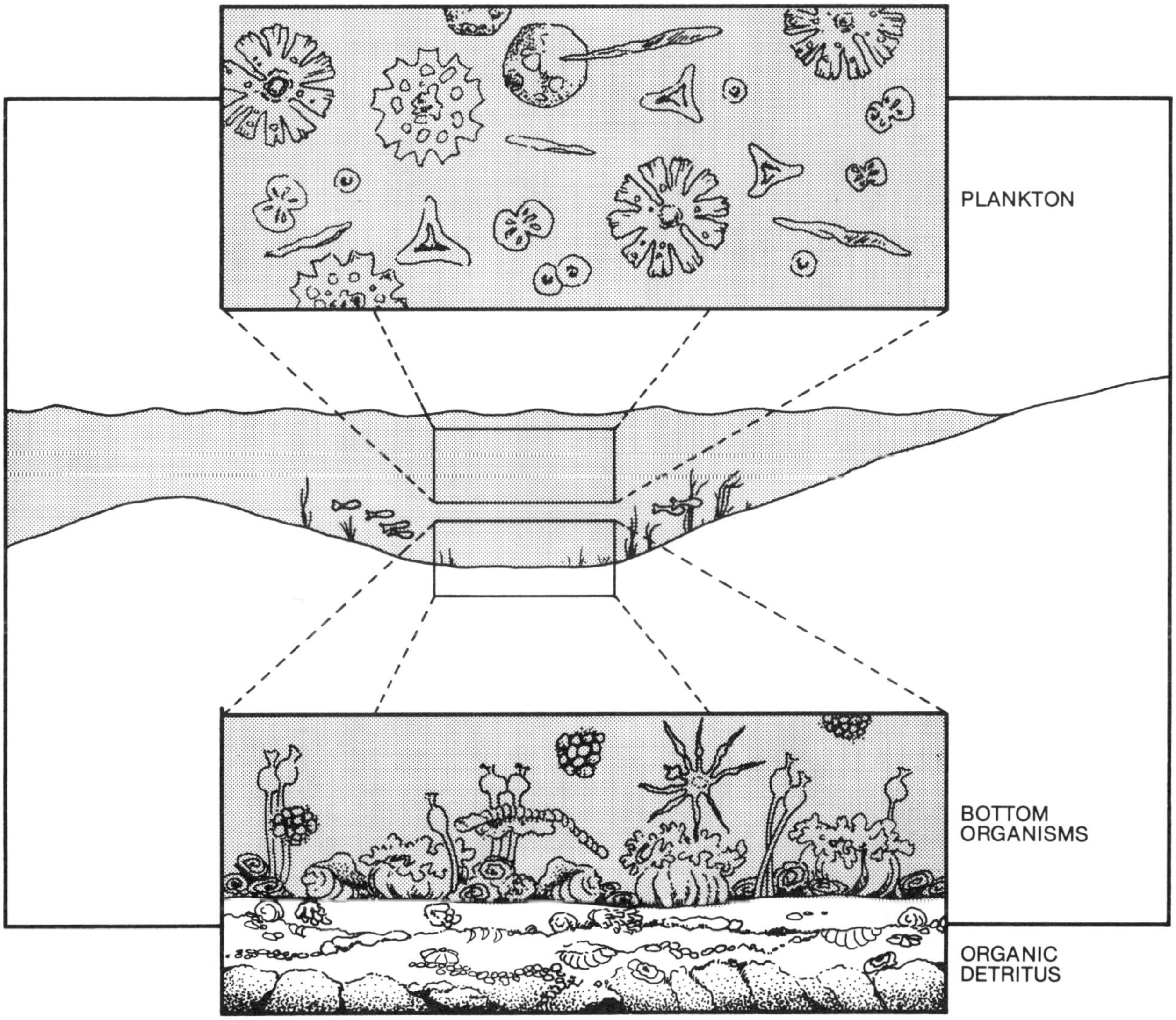

Figure 49. Shallow marine biotic community

the carbohydrates and proteins and leaving the carbon and hydrogen entombed in the mud.

As sedimentary beds build, the carbon- and hydrogen-rich mud is buried deeper and deeper and subjected to greater and greater pressures and temperatures. The clay and silt become hardened shale. The organic matter is trapped within its pores, where, under heat and pressure, it is chemically changed into hydrocarbons. At temperatures of 100° to 350° F, the hydrocarbons created are mostly oil. This process is especially efficient around 200° F, a typical formation temperature at a depth of 2 miles (fig. 50). At higher temperatures (at depths greater than about 4 miles) the mid-range carbon chains that constitute oil break down, leaving mostly gas with a carbon residue.

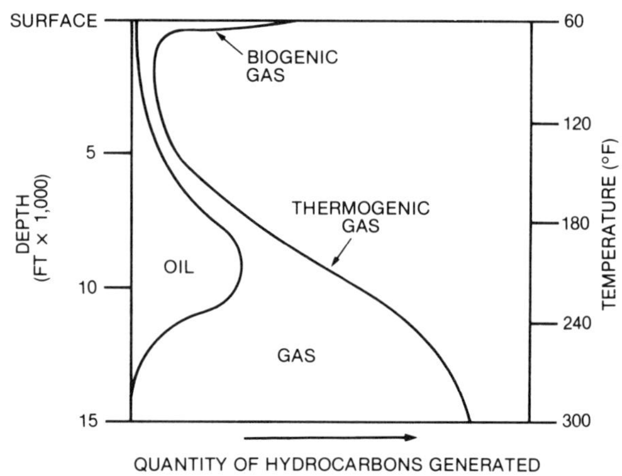

Figure 50. Relationship of burial depth to oil and gas generation

Migration

The movement of oil and gas out of its source rock and into reservoir rock is called *primary migration* (fig. 51). The pressure that results in the creation of hydrocarbons also forces fluids out of shale, which is easily compacted, and into rocks, such as sandstone, that retain more of their original porosity under pressure. Most porous rock, including shale, is saturated with fluids, particularly salt water trapped during the lithification of marine sediments. This connate (or *interstitial*) water carries oil and gas in suspension (and, to a minor extent, in solution) out of the compacted shale.

Porous rock is not necessarily permeable. To be *permeable*, rock must have pores that interconnect, allowing fluids to travel from one pore to another (fig. 52). Even though most shale is porous, it is relatively impermeable, because its pores are not connected very well.

Once inside the reservoir bed, oil and gas tend to rise to the top of the denser salt water. Buoyant oil globules rise more readily through the water as they coalesce into larger droplets. An accumulation of oil migrates upward through water until it encounters relatively impermeable rock (fig. 53). A bed of shale overlying reservoir rock thus forms a seal that prevents further upward migration and causes the oil to accumulate. Any further migration must be updip within the reservoir formation itself, either by the effects of buoyancy or by slow movement of formation water. This movement of hydrocarbons from one location to another within a reservoir bed, *secondary migration,* may involve distances up to many miles.

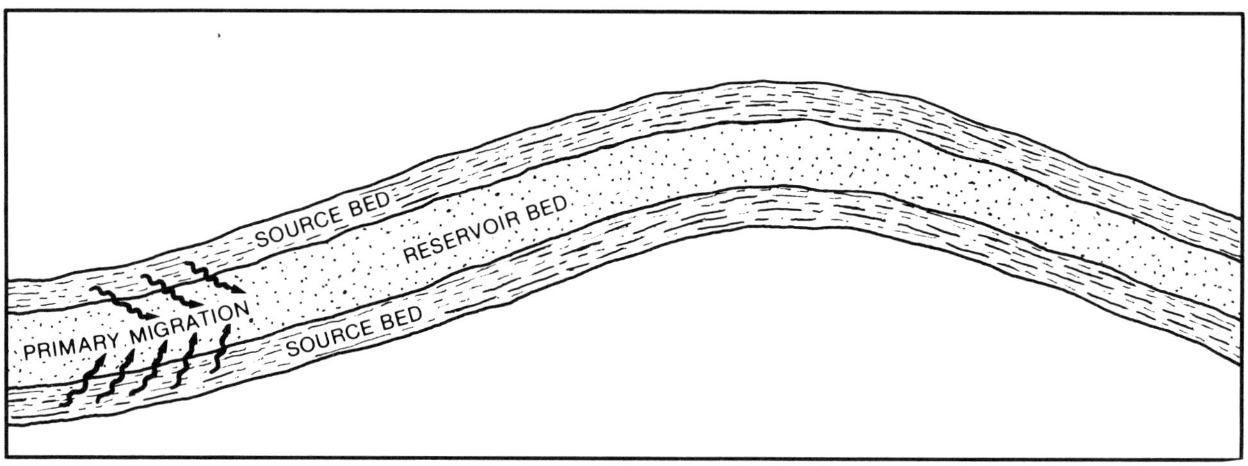

Figure 51. Primary hydrocarbon migration

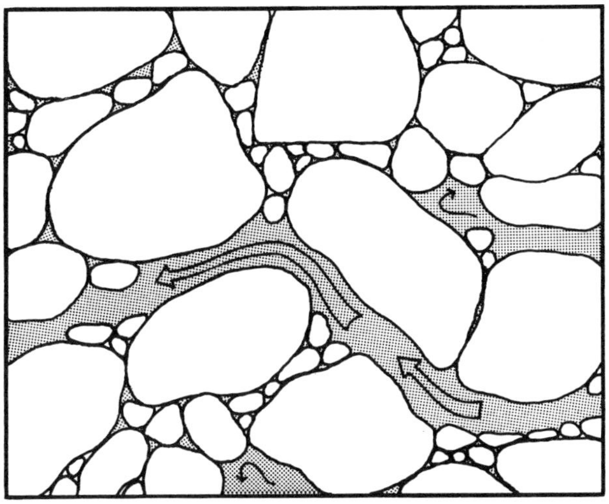

Medium Porosity, High Permeability

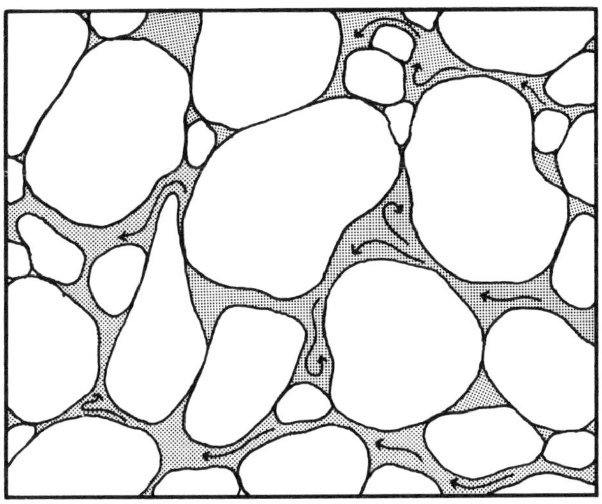

High Porosity, Low Permeability

Figure 52. Permeability versus porosity

Petroleum may continue to move updip until it reaches an outcrop or escapes through breaks in the overlying impermeable layers. If oil and gas reach the surface, they form a *seep* (fig. 54). The gas dissipates in the atmosphere, and the oil quickly decomposes, leaving behind only a tarry residue of the heaviest, least-volatile hydrocarbons. A large, though unknown, percentage of all oil and gas is lost this way.

Some of the migrating oil and gas, however, may reach a location beneath the surface where the configuration or physical properties of the rock make further upward movement impossible. Since the hydrocarbons can move into such a location but not out of it, they tend to accumulate. A location such as this—a *trap*—is what petroleum geologists look for as they collect data.

Traps

The accumulation of oil and gas in a trap may be the result of many factors acting together. If the principal trapping factor is the folding or faulting of the reservoir formation, the trap is said to be *structural*. If it is a lateral permeability change within the formation—such as cementation by minerals dissolved in groundwater—or the thinning out or disappearance of the reservoir formation updip, the trap is *stratigraphic*. If oil and gas accumulate in a given location because of the flow of formation water, the trap is *hydrodynamic*. Although these types of hydrocarbon traps are discussed separately here for purposes of clarity, most traps have both structural and stratigraphic features.

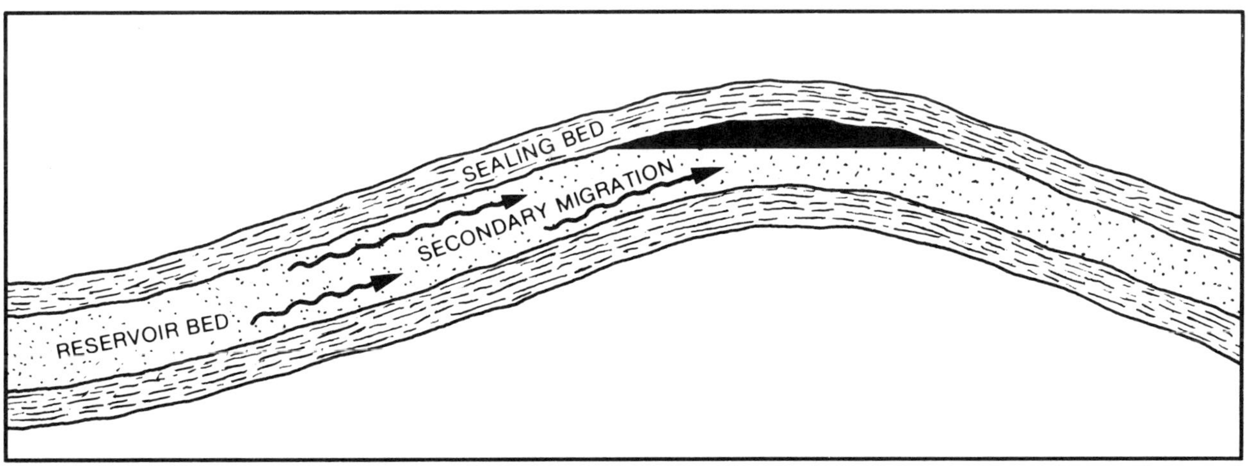

Figure 53. Secondary hydrocarbon migration

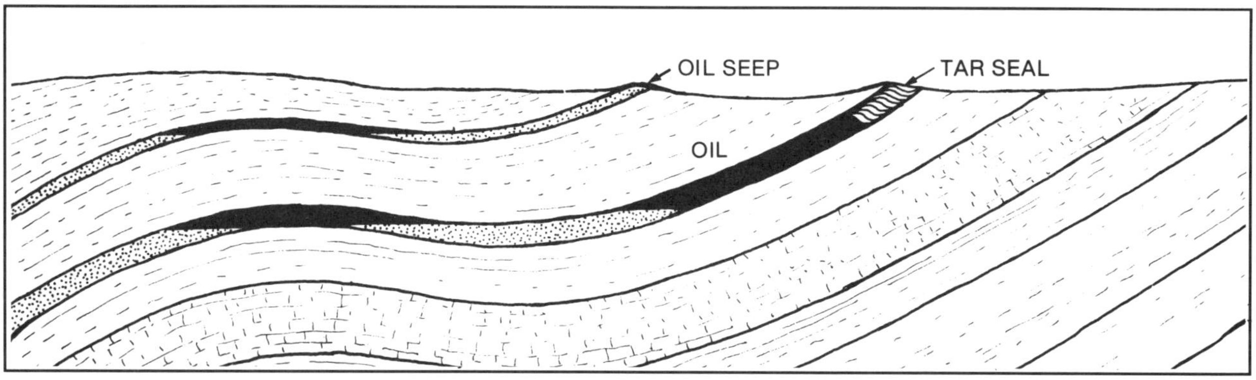

Figure 54. Hydrocarbon seeps

Structural Traps

Anticlines are the most common structural traps (fig. 55). They were the most commonly drilled reservoirs in Pennsylvania, where the modern oil industry began, because their surface outcrops were distinctive and easily recognized by drillers. Today much oil is produced from *salt domes,* inverted bowl-like upheavals often caused by rising columns of *halite* (rock salt), a type of short anticline common along the U.S. Gulf Coast.

An anticline usually consists of many rock layers of various porosities and permeabilities. Some of these layers may contain accumulations of oil and gas, while others form impermeable barriers. Wells drilled into such structures are often completed at several levels to produce from each reservoir formation.

Oil and gas are sometimes trapped by a fault crossing a reservoir bed (fig. 56). Petroleum migrating updip toward the fault is trapped by a layer of less-permeable rock that has been moved adjacent to the reservoir formation, or by a thin layer of pulverized rock *(gouge)* within the fault itself. Sometimes, however, a fault may serve as a permeable conduit through which oil and gas can escape from the reservoir. Folding and faulting often occur together, and anticlinal reservoirs often contain fault traps (fig. 57). A *growth fault,* in which deposition is greater on the downthrown side, can trap oil both along the fault and in its associated *rollover anticline* (fig. 58). Growth faults are common along the U.S. Gulf Coast, where deposition is rapid.

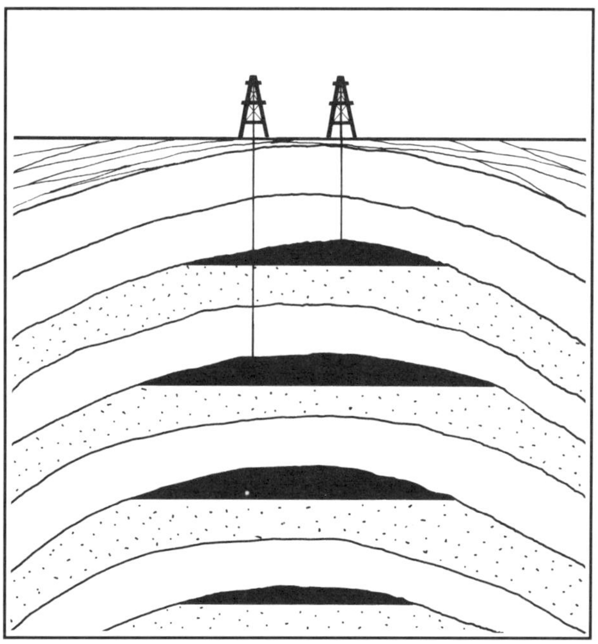

Figure 55. Anticline trap

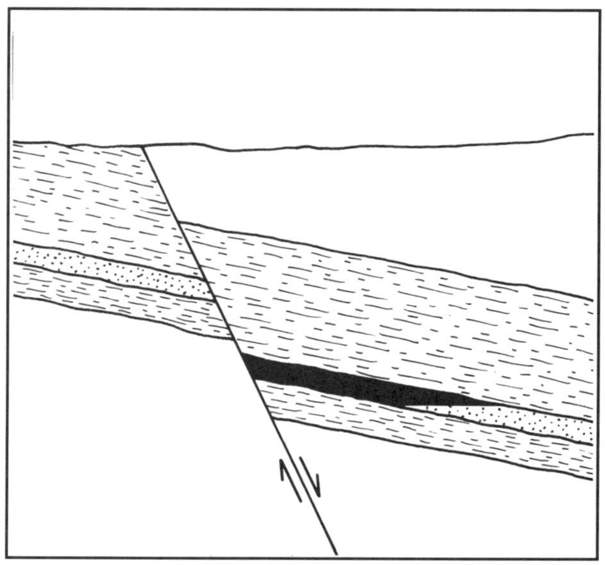

Figure 56. Fault trap

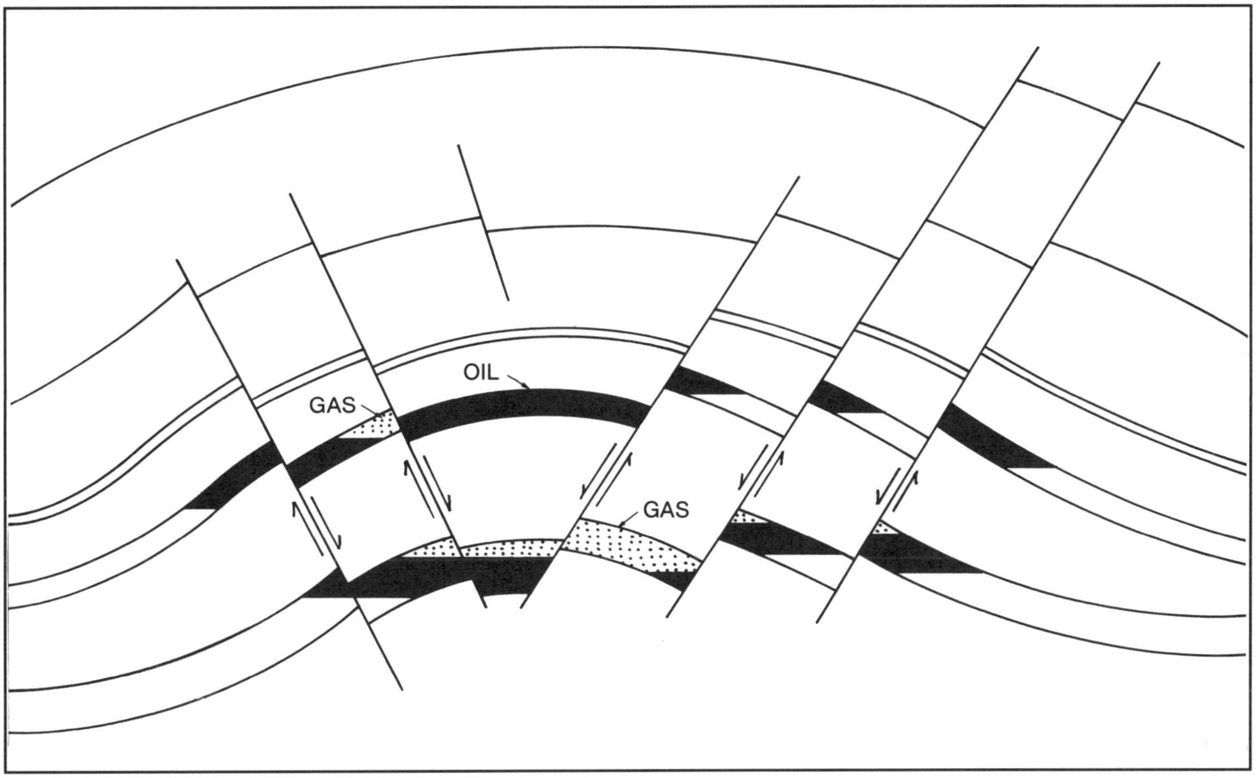

Figure 57. Faulted anticline traps

Stratigraphic Traps

One common type of stratigraphic trap is the *pinchout* (fig. 59*A*). Most rock layers thin out toward their edges. Oil or gas migrating toward the edge of a permeable reservoir formation may eventually be trapped where the formation "pinches out" between less-permeable rocks. A lateral decrease in permeability has much the same effect: even if the formation remains thick, the effective cross-sectional area of interconnected pores, through which oil and gas migrate, is reduced or eliminated (fig. 59*B*).

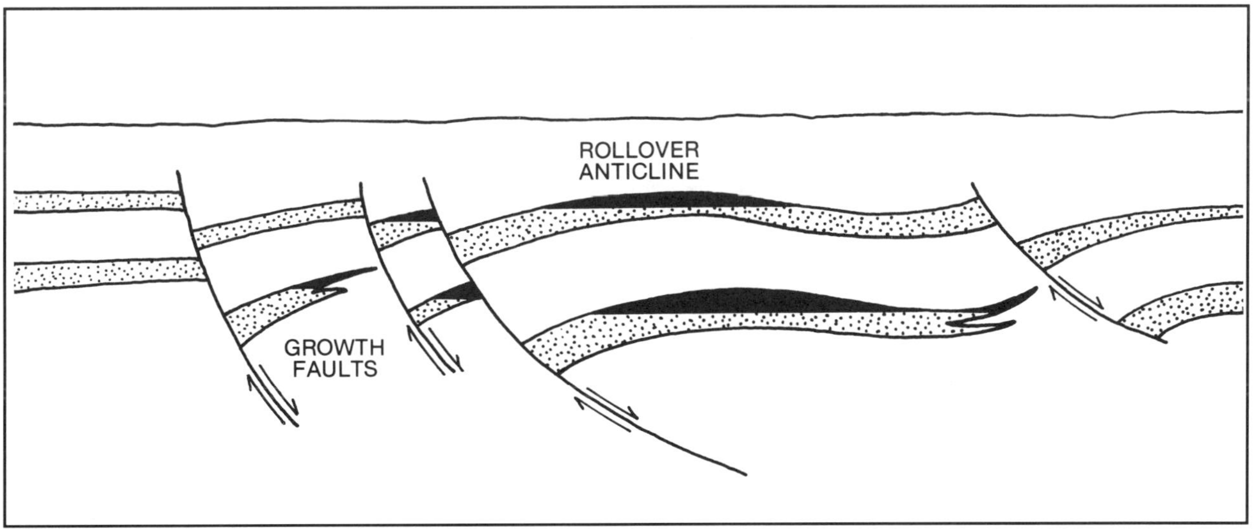

Figure 58. Rollover anticline and growth fault traps

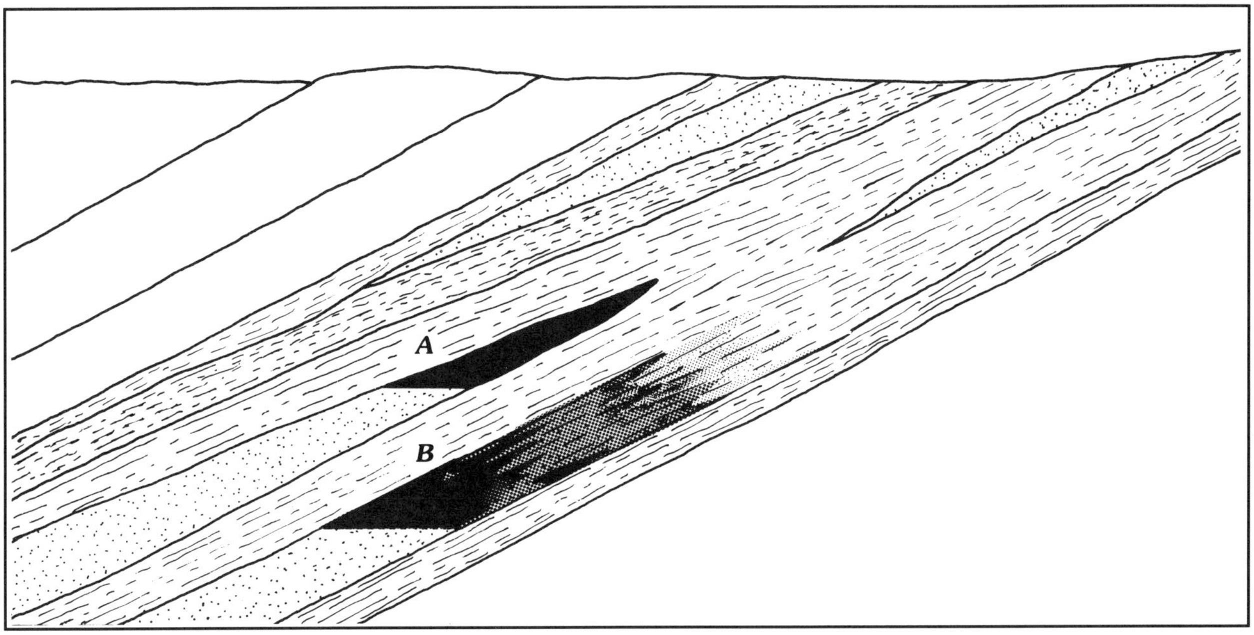

Figure 59. Stratigraphic traps: *A*, pinchout; *B*, lateral updip decrease in permeability

A *lens* is a relatively small body of sedimentary rock that pinches out in all directions between other formations (fig. 60). Sand lenses are often formed when a point bar on the inside of a river bend is cut off and buried beneath finer sediments in a backwater. Oil that migrates into a lens has no way to get out. A *shoestring sand* (fig. 61), essentially a connected series of sand lenses, is a long, sinuous sandstone reservoir that was formerly a sandy stream channel. A *bar sand* is a long, narrow sandstone formation made up of coastal beach sands or sandbars. A limestone *reef*—like a lens or bar sand, formed in place and isolated by other sediments—can serve as a hydrocarbon trap if sufficiently porous and permeable (fig. 62).

Another class of stratigraphic trap includes those associated with an unconformity (fig. 63). If a permeable outcrop, such as a sandstone ridge, is covered by a layer of clay that is later transformed into impermeable shale, hydrocarbons in the sandstone may be unable to migrate across the unconformity. Hydrocarbons generated in the shale, however, may be

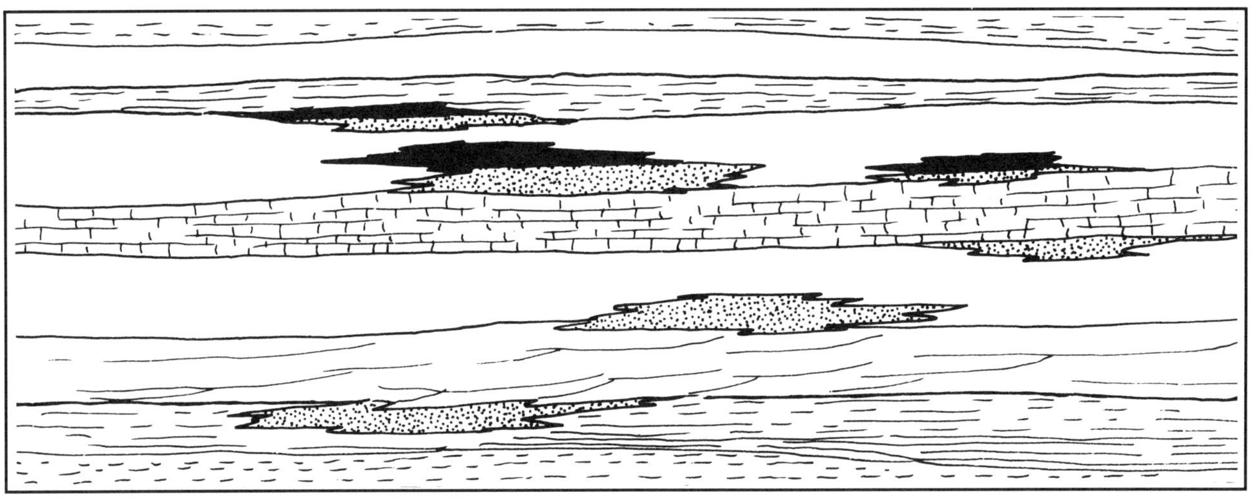

Figure 60. Lens traps

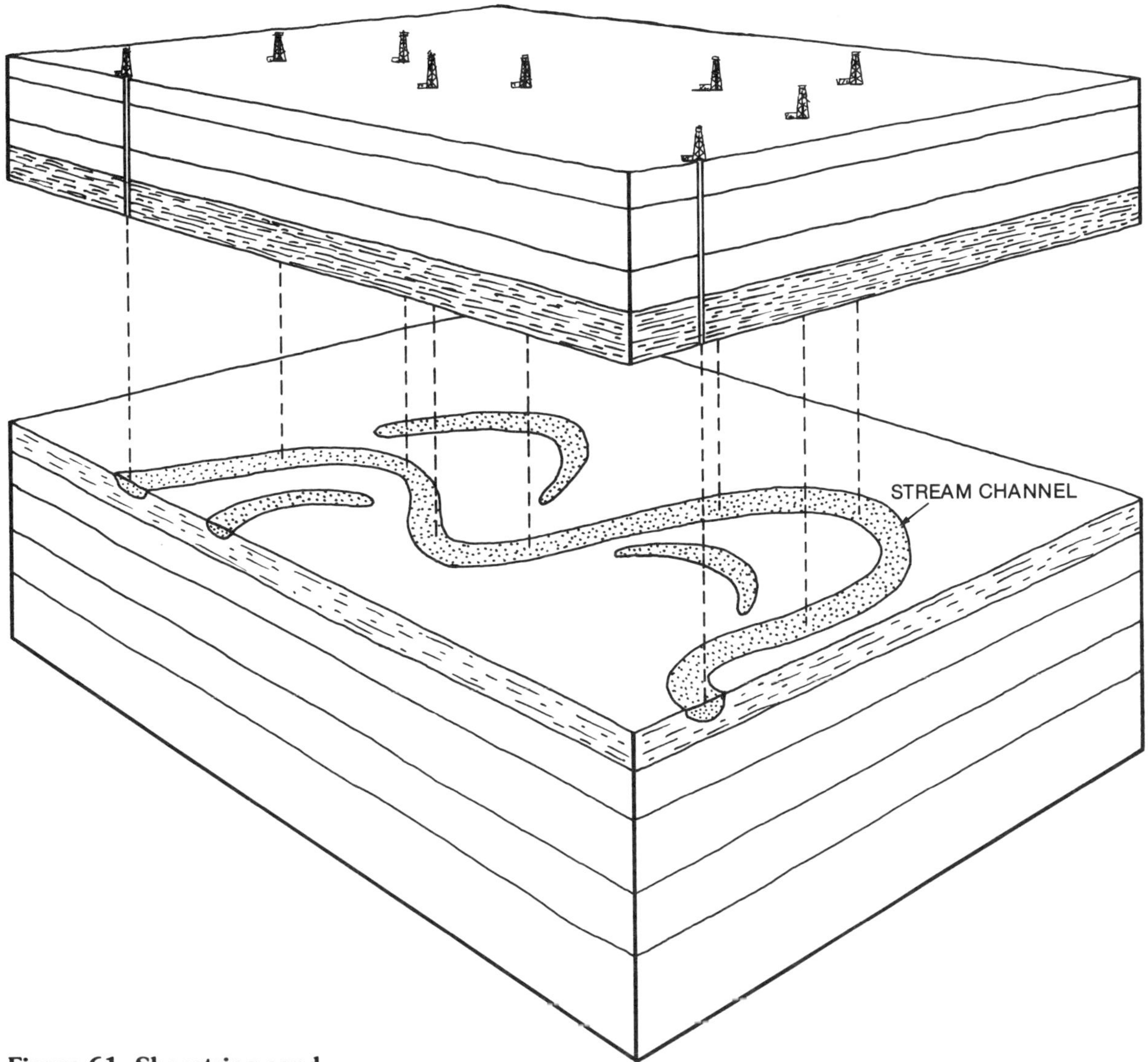

Figure 61. Shoestring sand

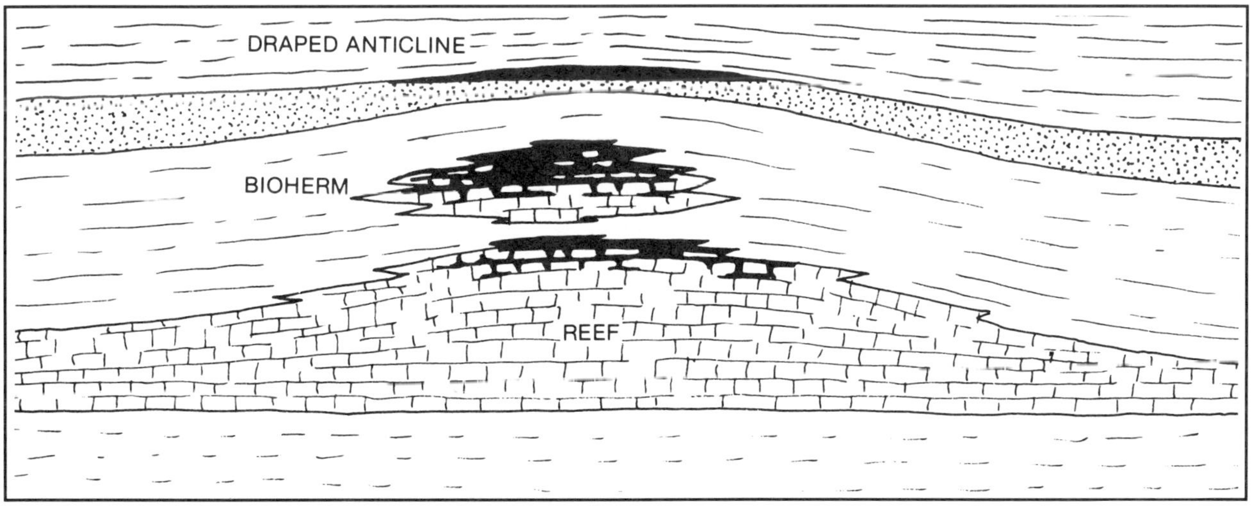

Figure 62. Traps associated with a limestone reef

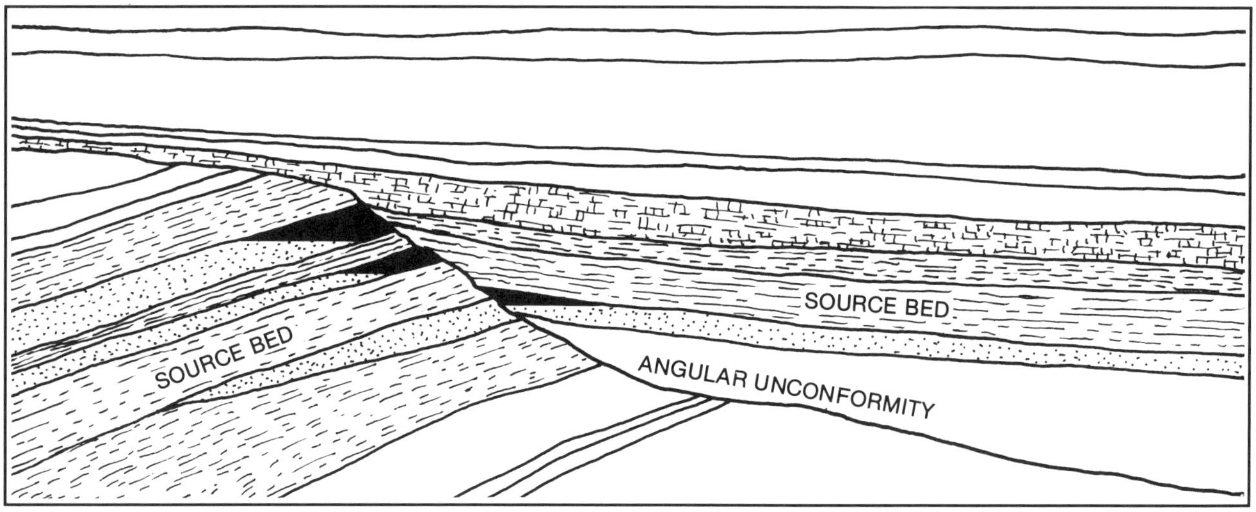

Figure 63. Hydrocarbon accumulation from source beds above and below an unconformity

forced into the underlying sandstone. Under some circumstances, oil may be trapped above the unconformity as well as below.

Some oilfields include both structural and stratigraphic traps in close proximity to one another. Along the U.S. Gulf Coast, layers of rock salt have been buried beneath thick beds of other sedimentary rock. The weight of the overlying rock has deformed the lighter, weaker salt, causing it to gather into a series of salt columns that rise slowly toward the surface. A *diapir* (or *piercement dome*) is a structure that is formed when the rising column penetrates and pushes aside the overlying layers (fig. 64). Because they are dissolved by groundwater, such salt columns rarely manage to reach the surface, but the layers that are domed upward or penetrated may contain many different kinds of traps.

The volume of hydrocarbon fluids a trap may contain is limited not only by the thickness and porosity of the reservoir formation but also by its *closure,* the total vertical height between the highest point in the reservoir and the *spill point,* the level at which trapped oil or gas can begin "spilling" upward and out of the trap (fig. 65). Closure is easy to see in a dome trap, which resembles an inverted cup. A cup can be filled only as high as the lowest point on the rim. If the cup is tilted, it can hold even less. The place where the water runs out of the cup is the "spill point," and the "closure" of the cup is the vertical difference in level between the spill point and the bottom of the cup. The spill point of a hydrocarbon trap is the lowest closed contour from which oil or gas can migrate out of the trap.

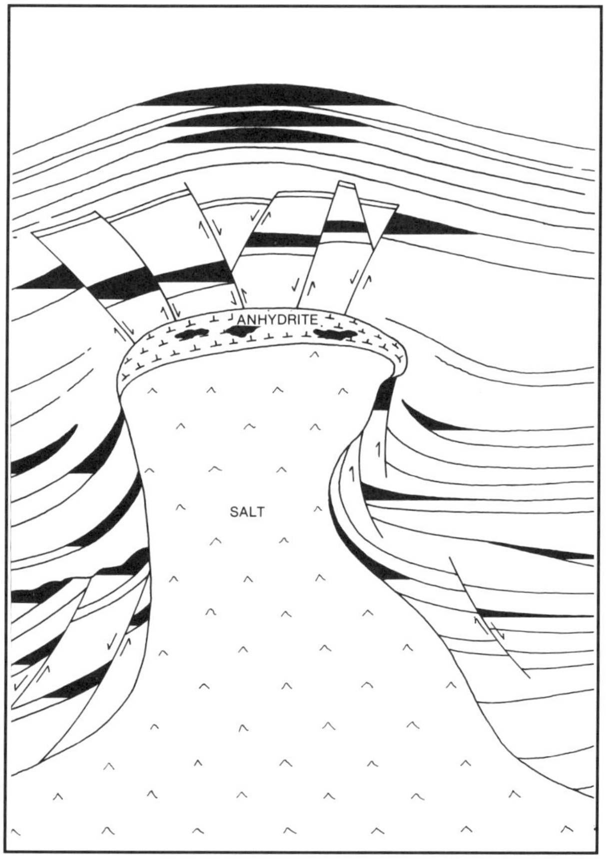

Figure 64. Hydrocarbon traps around a salt diapir

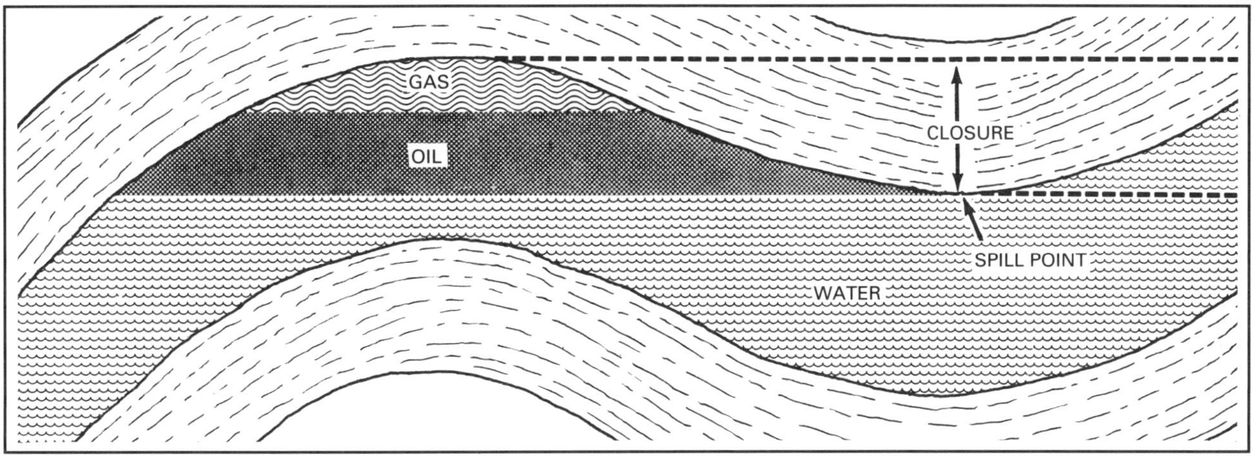

Figure 65. Closure in an anticline trap

Hydrodynamic Traps

In a stratigraphic or structural trap under hydrostatic conditions—with fluid at rest—the interface between water and oil is horizontal. *Hydrodynamic* trapping, by contrast, is the result of the slow flow of water through the reservoir formation in either a stratigraphic or a structural situation. Oil and gas are trapped by the dynamic pressure of the flowing water. In this type of trapping, the oil-water contact is tilted (fig. 66).

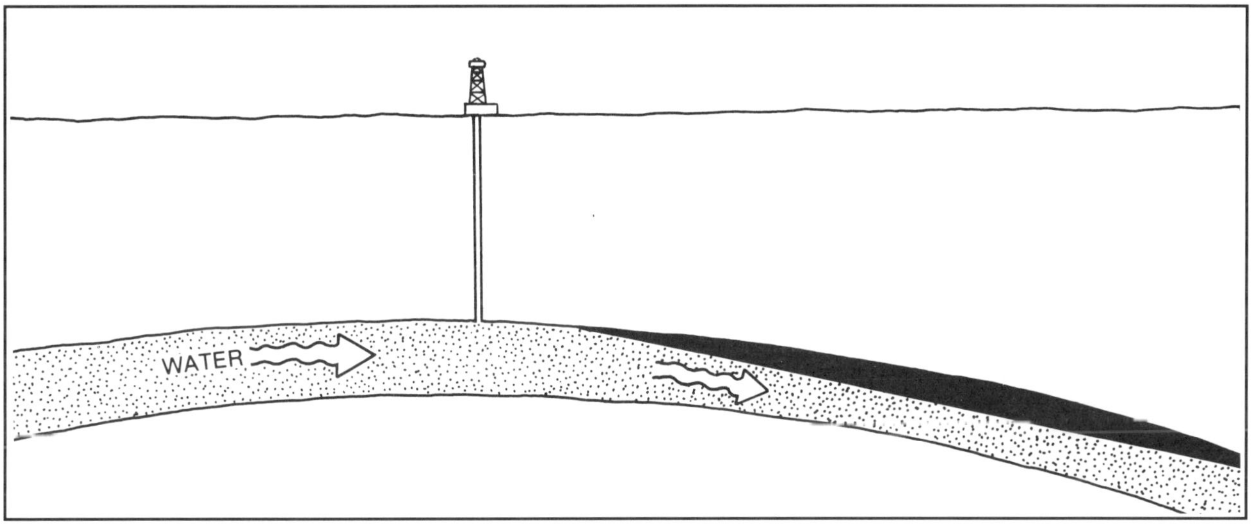

Figure 66. Hydrodynamically tilted oil-water contact

REVIEW QUESTIONS
Chapter 4

C 1. The chemistry of petroleum is based on the carbon chain molecule.

A. hydrogen
B. oxygen
C. carbon
D. nitrogen

C 2. The most common type of structural hydrocarbon trap is the—

A. pinchout.
B. fault.
C. anticline.
D. lens.

B 3. When organic matter is converted into hydrocarbons, oxygen is—

A. added.
B. removed.
C. not affected.

D 4. The structure formed when a salt column pushes upward through overlying formations is—

A. a syncline.
B. a shoestring sand.
C. a hydrodynamic trap.
D. a diapir.

T 5. (T/F) The lowest closed contour from which oil or gas can escape from a hydrocarbon trap is called the spill point.

C 6. The benzene molecule has ________ carbon atom(s).

A. one
B. three
C. six
D. eight

A 7. Secondary migration of hydrocarbons—

A. can involve distances up to many miles.
B. is caused by temperatures over 400° F.
C. is movement from permeable into impermeable rock.
D. always ends in a trap.

B 8. Hydrocarbons are generated from—

A. porous sandstone.
B. living organisms.
C. carbon dioxide and carbon monoxide.
D. nitrogen and water.

A 9. The source material for most petroleum is thought to accumulate in—

A. warm, shallow seawater.
B. dry grasslands.
C. granite.
D. the upper atmosphere.

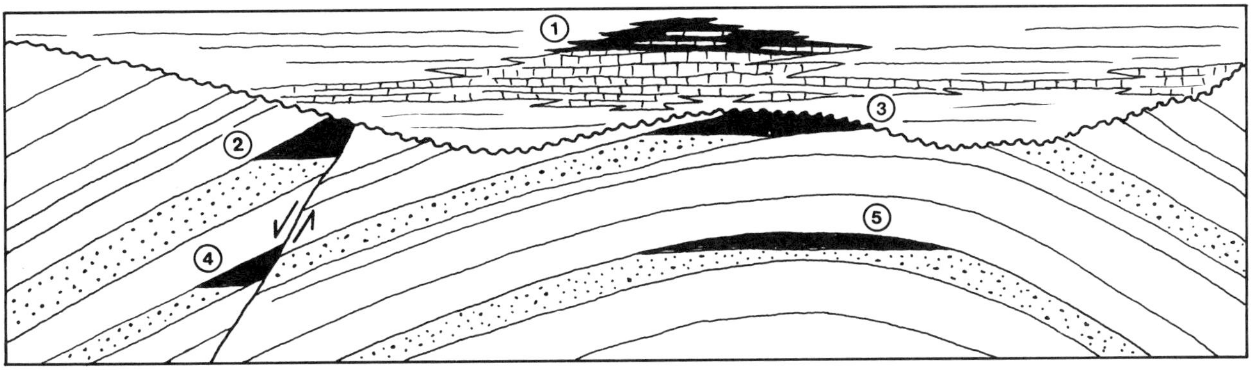

C 10. In the diagram above, hydrocarbon trap 1 is a _______ trap.

A. structural
B. hydrodynamic
C. stratigraphic

B 11. In the diagram, trap 2 is a ____________ trap.

A. structural
B. combination
C. stratigraphic

D 12. In the diagram, the purely structural traps are—

A. 1 and 3.
B. 1, 3, and 4.
C. 3 and 5.
D. 4 and 5.

B 13. In the diagram, hydrocarbons at 2 and 3 are trapped by—

A. faults.
B. an angular unconformity.
C. a nonconformity.
D. an anticline.

C 14. Oil is ________________ salt water.

A. denser than
B. as dense as
C. less dense than
D. never found near

T 15. (T/F) Oil and gas may be trapped above as well as below an unconformity.

CHAPTER 5
EXPLORATION

Introduction

Geologists have visited nearly every part of the planet in their search for petroleum. They are reasonably sure of which regions are most likely to have pools of oil or gas large enough, and easy enough to drill, to be commercially valuable. Recent advances in geology, especially in the study of plate tectonics, have helped geologists refine their predictions about where new discoveries are most likely to occur.

No oil-producing region has oil throughout; the resource is found only in particular locations. Oil pools are related to geologic features that are measured in feet, metres, kilometres, and miles. If rock were invisible, most oil "pools" would look like thin clouds of black smoke.

Since the earth is not transparent, the trick is to find the geologic layers and structures that are most likely to contain oil. Instead of seeing these features in the ordinary way, exploration geologists look into the subsurface by using energies to which the crust is somewhat transparent: gravitation, magnetism, and sound waves. To complete the subsurface picture in greater detail, they take samples and measure the properties of the rocks themselves, then compare rock sequences with those from other locations.

Geophysical Surveys

The two most-used geophysical exploration techniques are gravimetric and magnetic surveying. Although both methods give only a general idea of subsurface geology, they can cover large areas cheaply and are useful in narrowing the search for the most likely locations.

A *gravimetric survey* usually involves traversing a predetermined grid with a *gravimeter*, an instrument that can detect small local variations in the strength of the earth's gravity. These variations are caused by differences in rock density. Igneous and metamorphic rock are denser than sedimentary rock and cause a higher local gravity reading. The readings are mapped as contour lines connecting points of equal gravitational strength (fig. 67).

Most petroleum is found in thick sequences of sedimentary rock; oil rarely occurs in igneous or metamorphic rocks. To the explorationist, higher gravity means a thinner sequence of sedimentary layers atop the basement rock, and therefore less likelihood of an accumulation of petroleum.

The results of a magnetic survey somewhat resemble those of a gravimetric survey (fig. 68). Igneous and metamorphic rocks usually affect the earth's magnetic field more strongly than do sedimentary rocks. A magnetometer, carried in a truck or towed behind an airplane or a ship, is used to traverse a map grid to locate areas where basement rock is deeply buried beneath sedimentary rocks.

Seismic Surveys

After narrowing down the search to the most promising areas, the explorationist begins to use more precise techniques to examine the subsurface geology. The geophysical methods already described provide data on the depth of sedimentary rocks but little information on their type, thickness, porosity, or other characteristics. For a local, but more detailed, picture, the geologist obtains a seismic section.

A *seismic wave*, the shock wave generated by an earthquake or explosion, travels through rocks at different rates, depending on rock density, strength, porosity, and other factors. At the interface between rocks of different types, seismic waves are either reflected or refracted (bent), the same way that light waves are reflected or refracted going from air into water. The wave pattern that results depends on both the difference in sound velocity across the rock interface and the angle at which the shock wave encounters this velocity boundary.

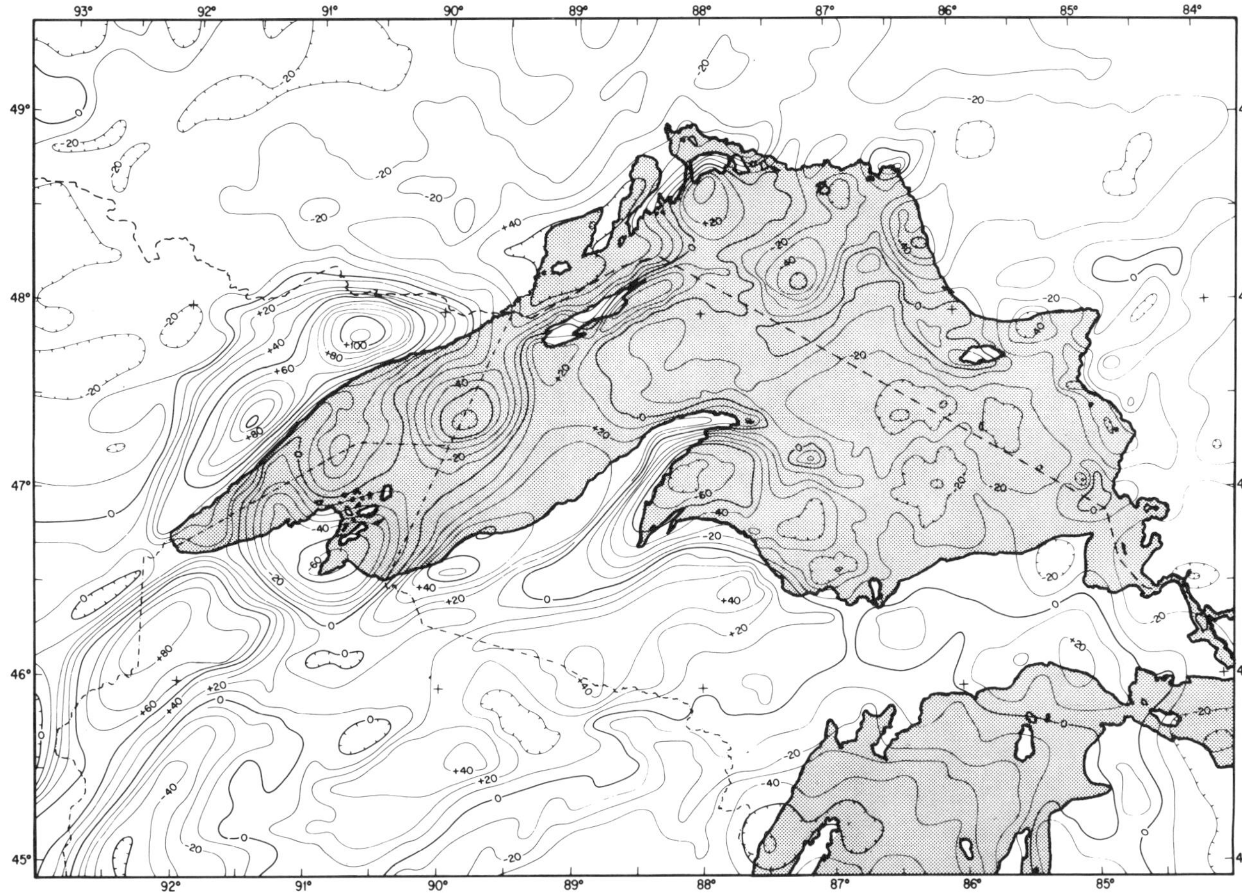

Figure 67. Gravity anomaly map of Lake Superior (Courtesy of United States Geological Survey)

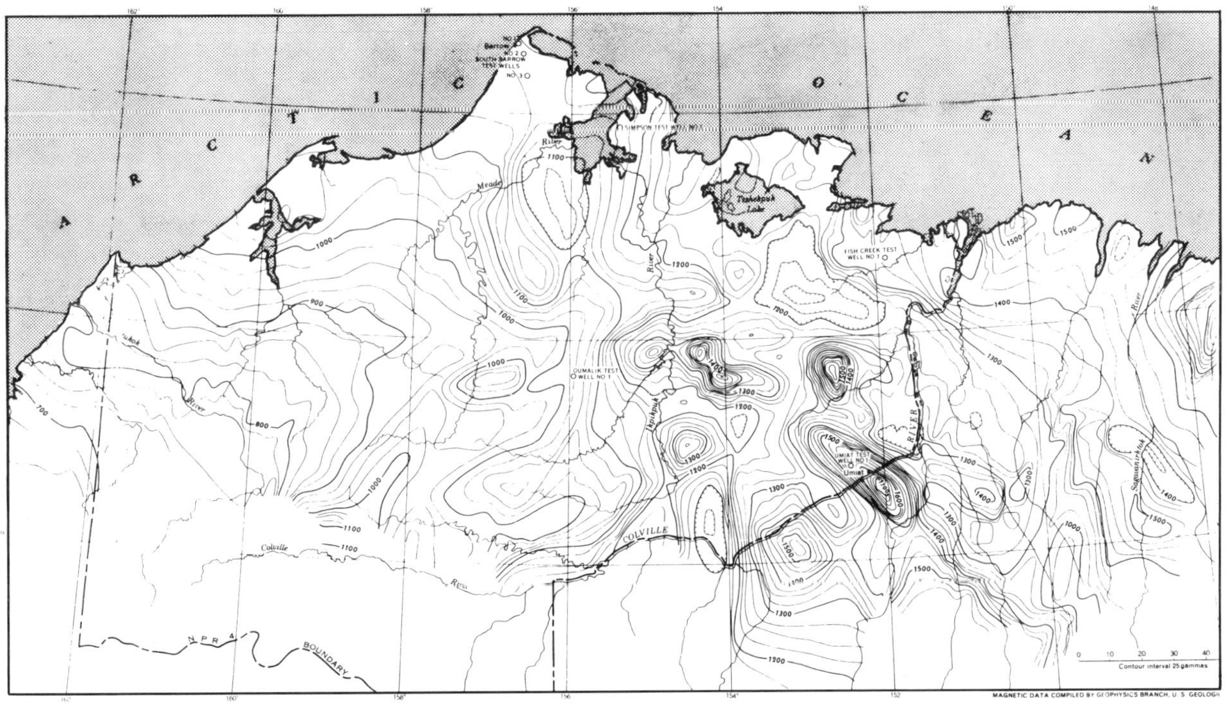

Figure 68. Magnetic contour map of northern Alaska (Courtesy of United States Geological Survey)

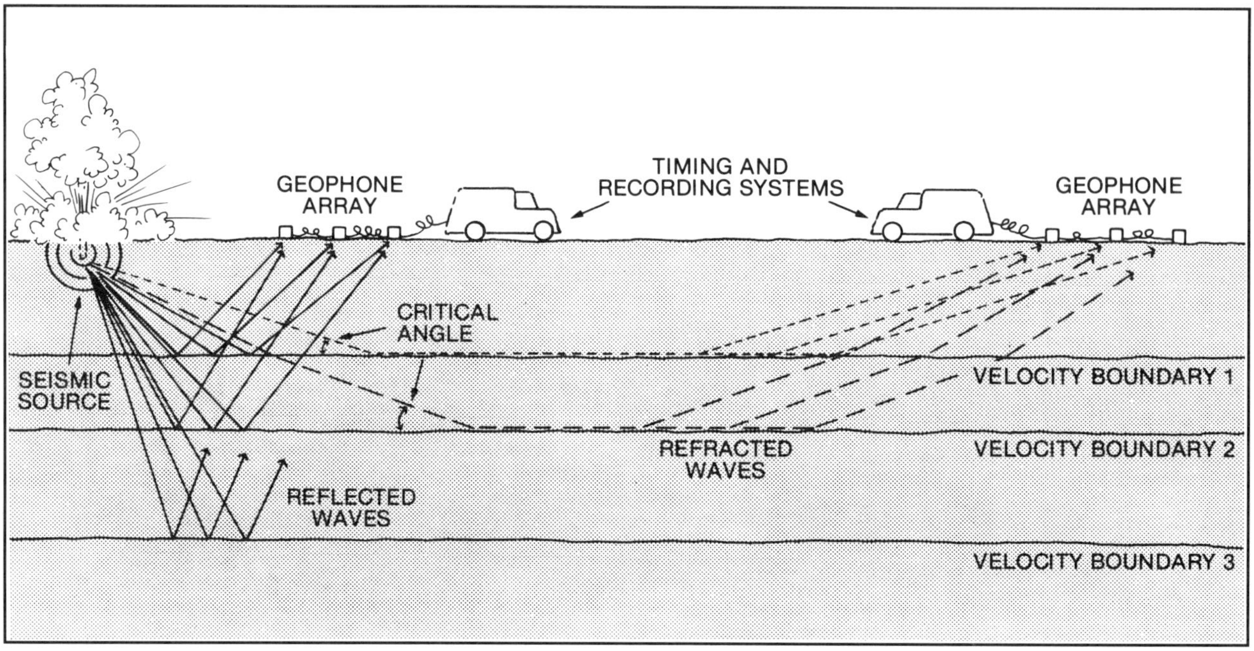

Figure 69. Seismic ray paths

In a seismic reflection survey, a shock wave is generated at the surface, either by setting off an explosion in a shallow hole or by using a mechanical wave source at the surface (fig. 69). The time it takes for a shock wave to travel from the surface, reflect off a velocity boundary, and return to each of several detectors *(geophones)* in an array is measured. A computer is used to analyze a digital recording of the return and generate a *seismic profile* or *record section* (fig. 70). The waves rebounding from different horizons are sorted out and appear as lines defining the upper and lower boundaries of different layers. Often the geologist already has an idea of what formations will be encountered; the seismic profile simply shows the depths at which these formations occur and any structure, such as folds or faults, that could trap oil and gas.

A seismic refractive survey covers a larger area, but to a lesser depth and with less resolution, or precision. Seismic waves that strike a velocity boundary at the *critical angle* are neither transmitted through the lower layer nor reflected back into the upper layer; instead, they travel along the boundary, so that their energy may be detected a considerable distance away. In some marine refraction surveys, geophones are planted on the seafloor to detect pulses generated up to 50 miles away.

Strat Tests

Exploration does not end when the first hole is drilled. The information about rock types and formation depths that can be obtained during the drilling of a well, whether a wildcat or a development well, adds detail to the subsurface picture that the geologist puts together. During the drilling of any hole, the bit grinds rock into fragments that are circulated to the surface in the drilling fluid. These *cuttings* are analyzed and identified by the mud logger or the geologist to determine what formations or rock types are being penetrated (fig. 71).

Even before the first wildcat, however, a test hole called a *strat test* is sometimes drilled solely to gather geologic data. Because no oil is to be produced from it, a strat test is smaller in diameter than either an exploratory well or a development well. It can be drilled in less time, and with less expensive equipment, than a larger hole.

Coring is one of the sampling techniques used in drilling both strat tests and larger holes. A special bit is installed to cut out and recover a cylinder of rock 2 to 5 inches in diameter and 30 to 60 feet long (fig. 72). This technique provides a relatively undisturbed sample of the rocks in the same order that they occur beneath the surface.

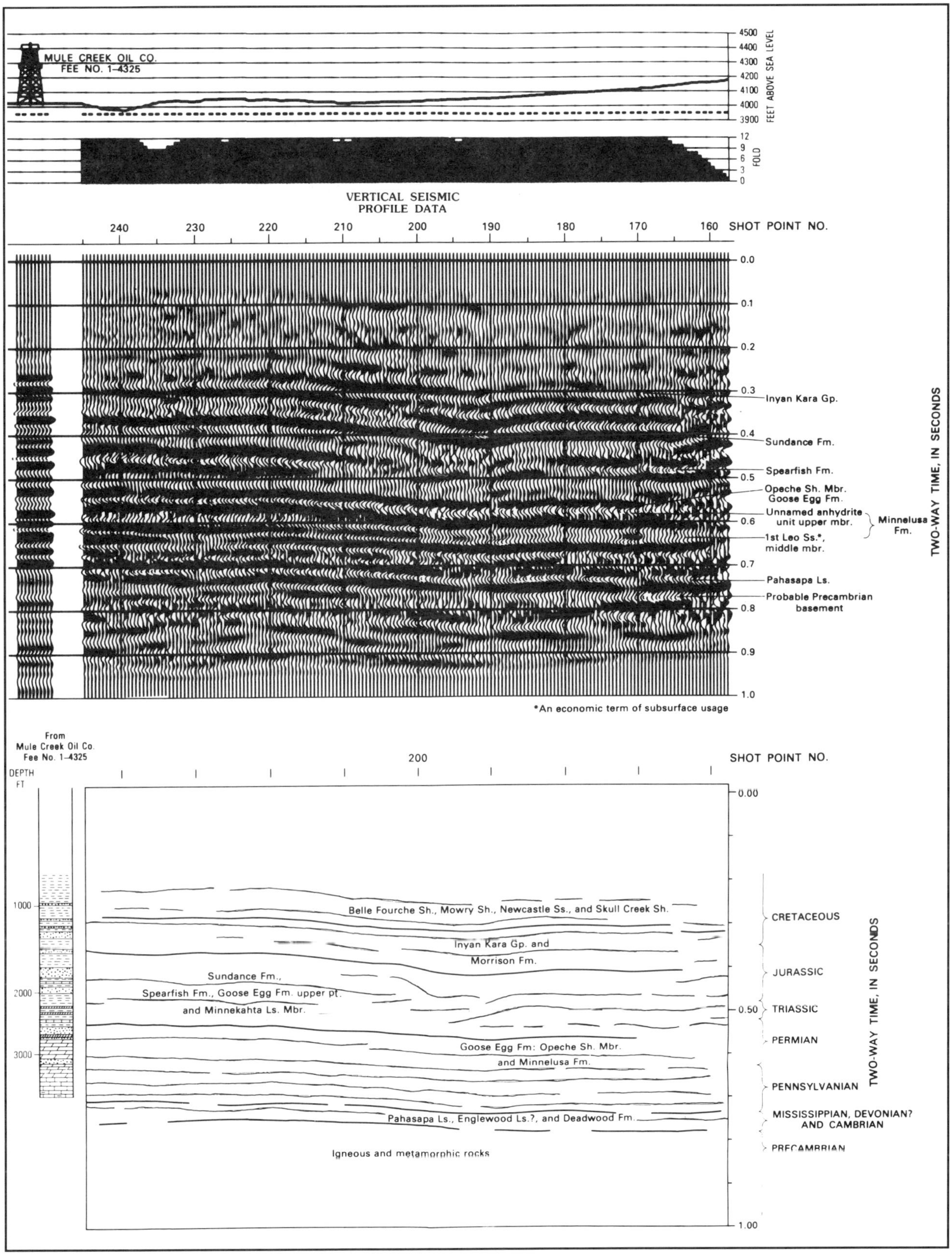

Figure 70. Seismic record section (Courtesy of Lee et al., United States Geological Survey)

Figure 71. Cuttings

Logging

During or following the drilling of a hole, wireline logs are used to measure formation characteristics at various depths. An instrument package, a *sonde*, is run into the hole on electric wireline (fig. 73). As the sonde is slowly retrieved, it transmits signals to the surface, where they are recorded in digital form for later computer analysis.

One of the simpler types of logs is the *caliper log* (fig. 74). This instrument mechanically measures hole diameter, which varies with depth because some rock types are prone to crumble and slough into the hole. The diameter of the hole also affects the readings obtained when using other types of wireline logs (fig. 75).

A *spontaneous potential* (*SP*, or *self-potential*) *log* measures the natural electrical charge that is induced when two liquids of different salinities come

Figure 72. Core sample (Photo by T. Gregston)

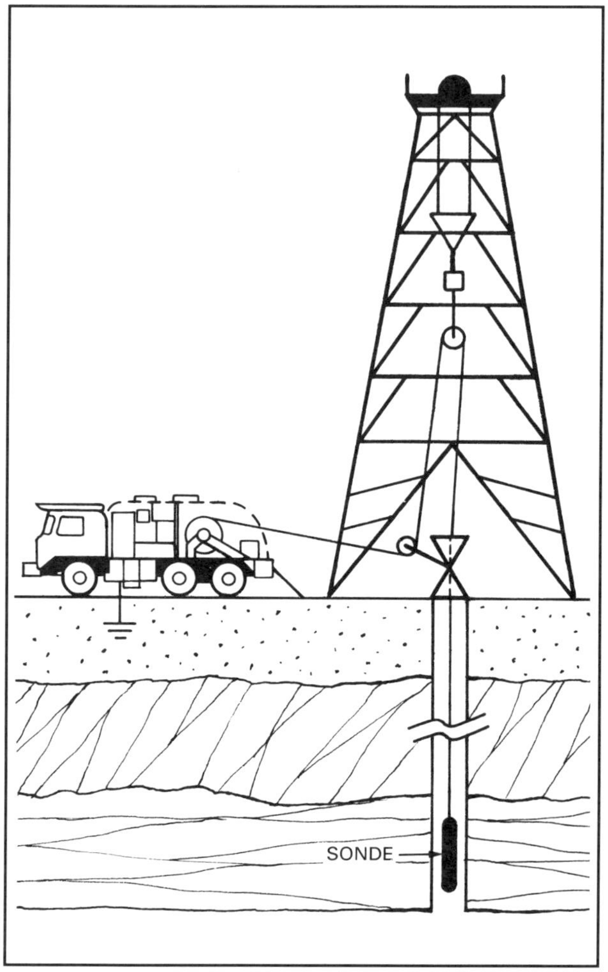

Figure 73. Lowering a logging sonde into the hole

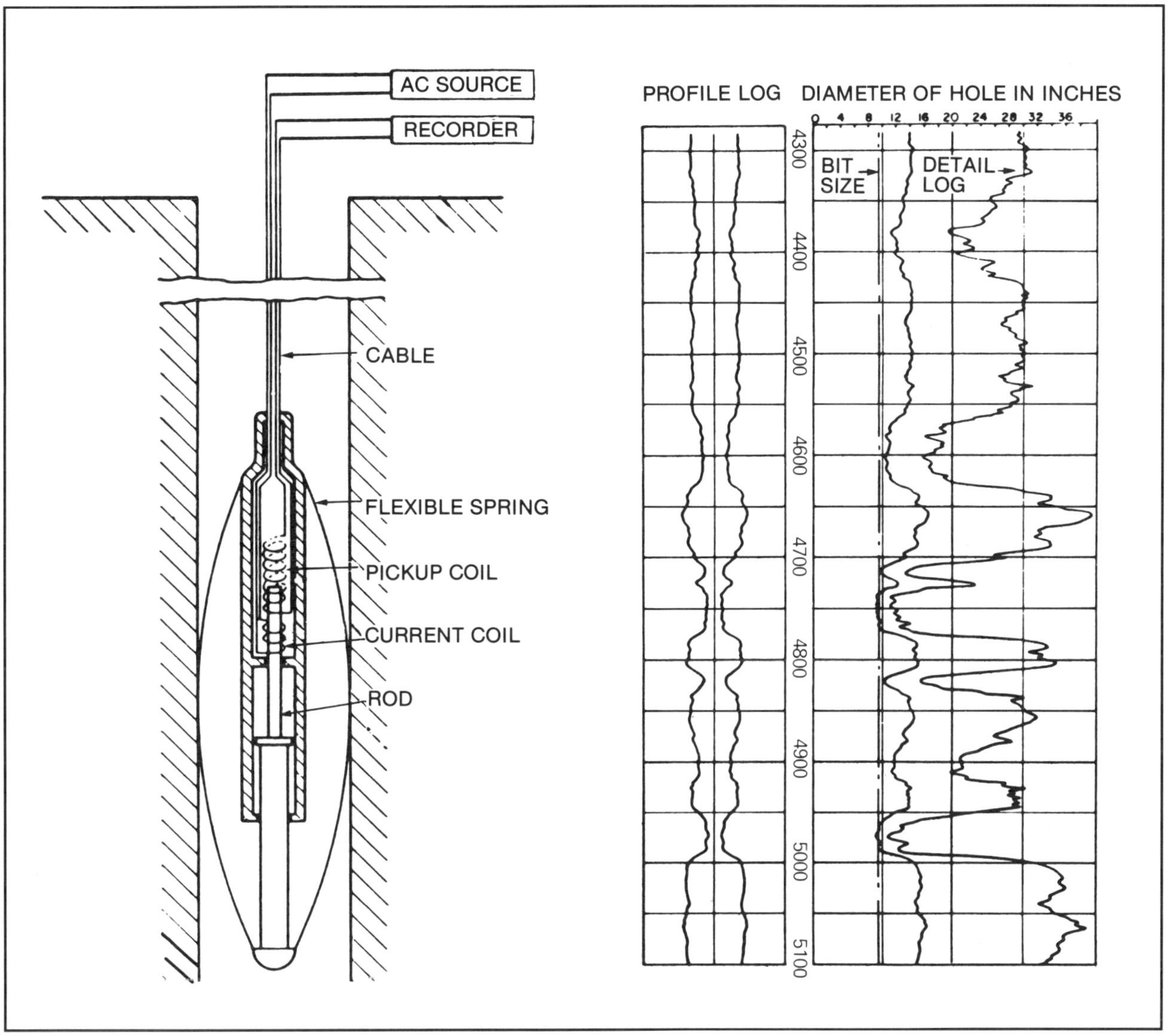

Figure 74. Caliper device and caliper log presentation

into contact. When freshwater drilling fluid invades a permeable formation, a spontaneous potential is created between that formation and an adjacent formation that has not been invaded but that contains saline formation fluids. The SP log is thus a good indicator of the presence of a permeable rock layer (fig. 75).

A *resistivity log* measures formation reaction to an electrical current induced by the logging instrument. The amount of current that will flow through the formation between two fixed points is measured. This current is inversely proportional to the electrical resistance of the formation, which depends on total porosity, pore shape and arrangement, salinity of formation water, and the amount of oil or gas present (fig. 75).

Radioactivity logs measure both natural and induced low-level radiation characteristics of formations. The *gamma ray log* detects high-energy radiation from naturally occurring radioactive elements that are more common in shale than in other types of rock. The *neutron log* bombards formations with radiation and records the amount of radiation that is not absorbed, an indirect measure of the hydrogen ion content of the formation, and thus of porosity (fig. 75).

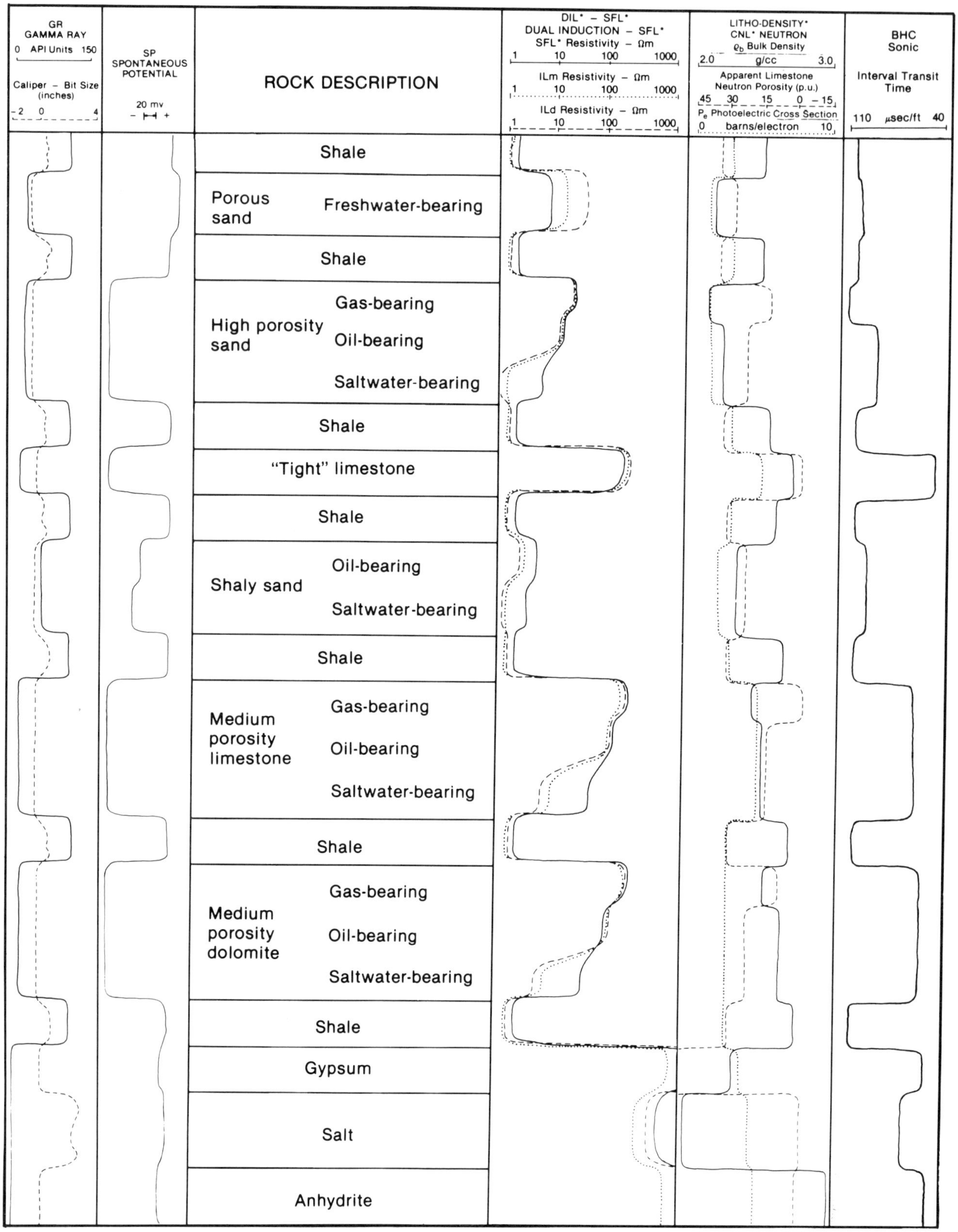

*Mark of Schlumberger

Figure 75. Composite log chart (Courtesy of Lewis Raymer, Schlumberger Well Services)

Correlation

Geologists trace the extent of a rock layer *(stratum)* over a broad area by looking for exposures in different places. Most rock layers vary in character and detail from one place to another due to differences in depositional environment. Often a geologist can identify a particular layer only by its relationship with other layers. For example, formation *B* in figure 76 grades from limy sandstone in the cliff on the left to shaly limestone in the canyon on the right. With no other information available, the geologist could assume that these outcrops were of different formations. However, a particular kind of fossil occurs at the top of the layer in both places; moreover, formations *A* and *C* are known to be continuous from one location to another. The geologist concludes that formation *B* represents the same geologic time in both areas even though the depositional environment varied from place to place.

Natural outcrops, such as in canyon walls, are convenient places to look for exposures for correlation. For most rock layers, however, and in most areas that interest petroleum geologists, such exposures are rare. The geologist usually relies on cuttings, core samples, or wireline logs obtained from wells or strat tests. Formations can then be correlated in an area selected specifically for its petroleum potential.

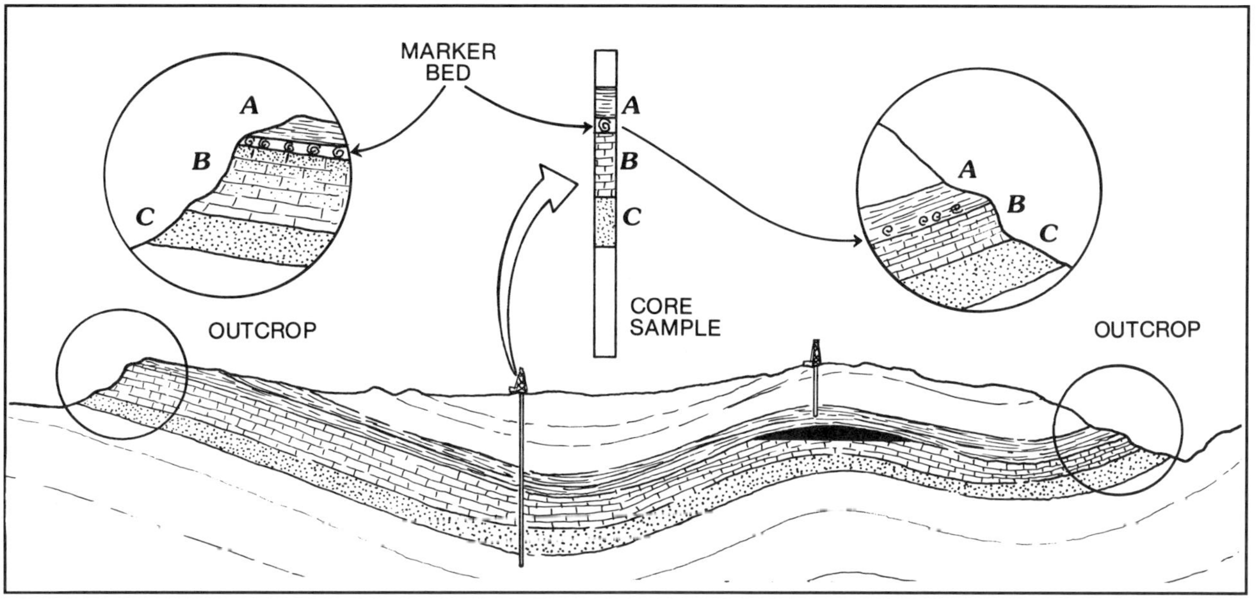

Figure 76. Correlation

REVIEW QUESTIONS

Chapter 5

____ 1. Which of the following geologic exploration techniques provides the most detailed information about rock types?

- A. gravimetric surveying
- B. magnetic surveying
- C. seismic surveying
- D. wireline logging

____ 2. (T/F) Core samples can be obtained during the drilling of a development well.

3. The matching of rock types and sequences from one area to another is known as ____________.

____ 4. Which of the following provides the best information about the sequence of rock types beneath the surface?

- A. a magnetometer
- B. cuttings
- C. a core sample
- D. a caliper log

____ 5. In seismic surveying, instruments detect and record shock waves reflected from—

- A. boundaries between rock layers.
- B. high-density fluids.
- C. large rock grains.
- D. oil and gas.

6. A ____________ log measures hole diameter.

____ 7. What is the main advantage of gravimetric and magnetic surveying?

- A. They pinpoint fault traps.
- B. They cover large areas at minimum cost.
- C. They distinguish between sedimentary rock types.
- D. They can detect gas in the atmosphere above an oil pool.

____ 8. A locally high gravity reading indicates—

- A. a thick sequence of sedimentary rocks.
- B. the presence of oil or gas.
- C. igneous or metamorphic rock at shallow depth.
- D. water-saturated sandstone.

____ 9. (T/F) A magnetic survey cannot be conducted over water.

____ 10. When seismic waves strike the boundary between rock types at the ____________, they are transmitted along it.

- A. critical angle
- B. velocity boundary
- C. record section
- D. water table

B 11. The results of a magnetic survey indicate—

A. accumulations of natural gas.
B. the depth of the basement rock.
C. velocity boundaries between rock types.
D. differences in rock density.

D 12. Which of the following logs measures natural formation radiation?

A. neutron log
B. spontaneous potential log
C. caliper log
D. gamma ray log

13. An instrument used in seismic surveying to detect reflected pulses is called a(n) geophone.

T 14. (T/F) The hole drilled for a strat test is smaller in diameter than that for a wildcat well.

C 15. A spontaneous potential log measures—

A. the resistance of a formation to an induced electric current.
B. the rate of flow of formation fluids.
C. the electric charge between fresh water and salt water.
D. natural formation radioactivity.

CHAPTER 6
DRILLING THE WELL

Introduction

The characteristics of the formations that must be penetrated to reach the target, whether drilling a wildcat in an unknown area or a development well in a proven field, affect the speed and efficiency with which the hole can be drilled. To be economically feasible, the hole must be drilled as directly as possible to the target. It should not delay the process by caving in and trapping the drill string or by sticking the pipe to the hole wall so that an expensive fishing operation becomes necessary. It must not empty its drilling fluid into highly permeable formations, nor take on substantial amounts of fluids from high-pressure formations—either case could result in a blowout. It should, however, be drilled with the exact combination of drill bit, bottomhole assembly, weight on bit, rotary speed, hydraulics, and mud type that makes hole with the smallest expenditure of equipment, time, and effort.

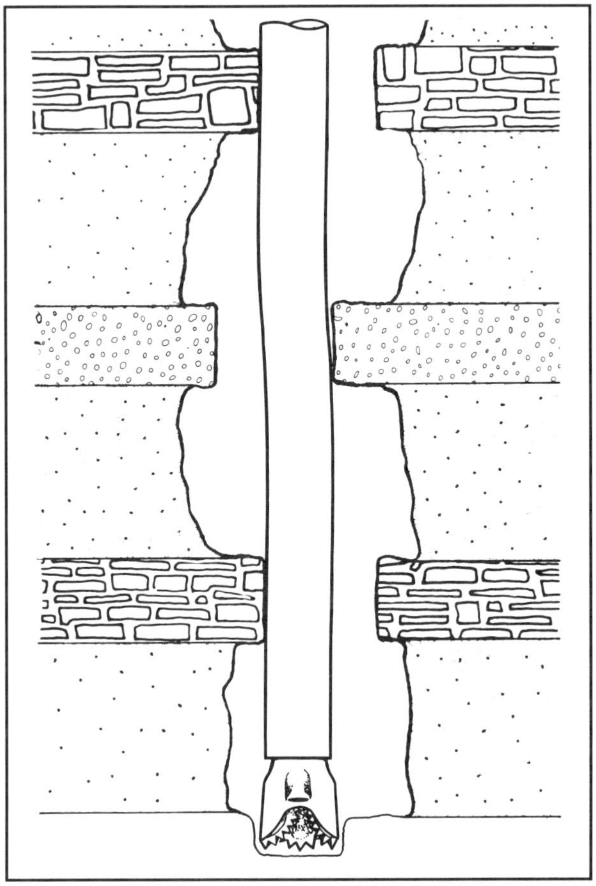

Figure 77. Variations in hole diameter caused by differences in sedimentary rock layers

Crooked Hole

Drilling through a series of sedimentary layers is complicated by variations in formation properties and their effects on the drill stem and bit. A drill string a few inches in diameter and hundreds or thousands of feet long is as limber as a strand of uncooked spaghetti. How well the driller can make the bit drill where he or she wants it to drill depends largely on the condition of the entire length of the hole and the characteristics of the rock that the bit is drilling through. If the layers are horizontal but of different strength, hardness, and integrity, the bit tends to drill out a crooked hole of varying diameter (fig. 77). If the driller knows of this tendency in advance, he or she can compensate for it with a particular *bottomhole assembly*—the combination of drill collars and stabilizers above the bit—to keep the hole on course.

If the formations being drilled are not horizontal, the bit tends to drift sideways, deflecting the hole away from the vertical. In formations slanted less than about 45°, the bit tends to drill *updip*—that is, perpendicular to (straight across) the formation (fig. 78*A*). Formations tilted more than 45° cause the bit to drill *downdip*, parallel to the layers, and to "walk" to the right in the formation plane (fig. 78*B*). The driller must counter these tendencies early by adjusting weight on bit and rotary speed or by using the appropriate bottomhole assembly. If the hole gets too far off course, the driller may have to run in special equipment to correct hole angle or redrill the hole below a certain point, adding enormously to the cost.

Sloughing or Swelling Shale

Some shales absorb water from the drilling fluid, expand, break away, and fall downhole. Large masses may lodge around drill collars and stabilizers, sticking the drill string and blocking circulation (fig. 79). Abnormally pressured shale, steeply dipping shale beds, and erosion by circulating drilling fluid can also cause the hole to cave in. The driller may adjust the *reactive solids content* of the mud—the amount of water-absorbent material—or add other chemicals to reduce water loss to a shale formation. The driller may also weight up the mud to increase pressure against the formation.

If the shale swells but does not slough off, the swollen layer may close around the drill pipe, cutting off circulation and trapping tool joints or drill collars (fig. 80). Using reamers in the drill string helps keep the hole open and equipment free.

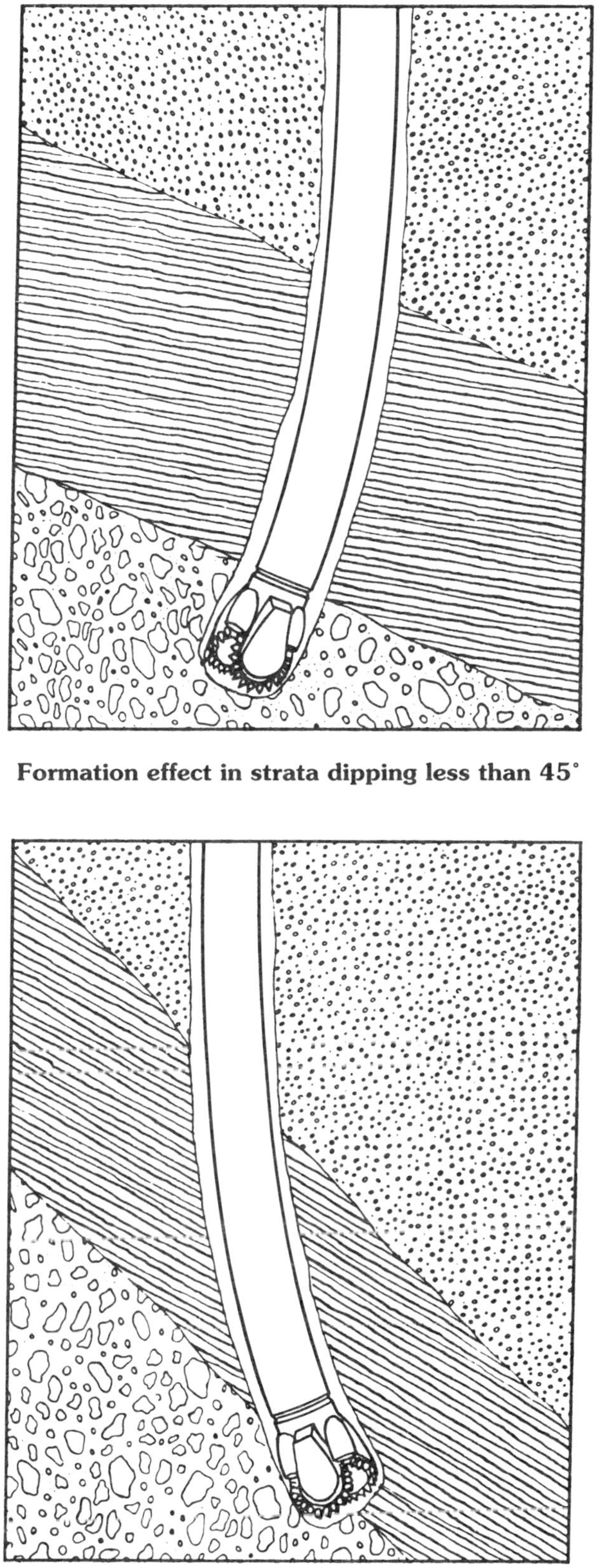

Formation effect in strata dipping less than 45°

Formation effect in strata dipping more than 45°

Figure 78. Bit drift in nonhorizontal formations

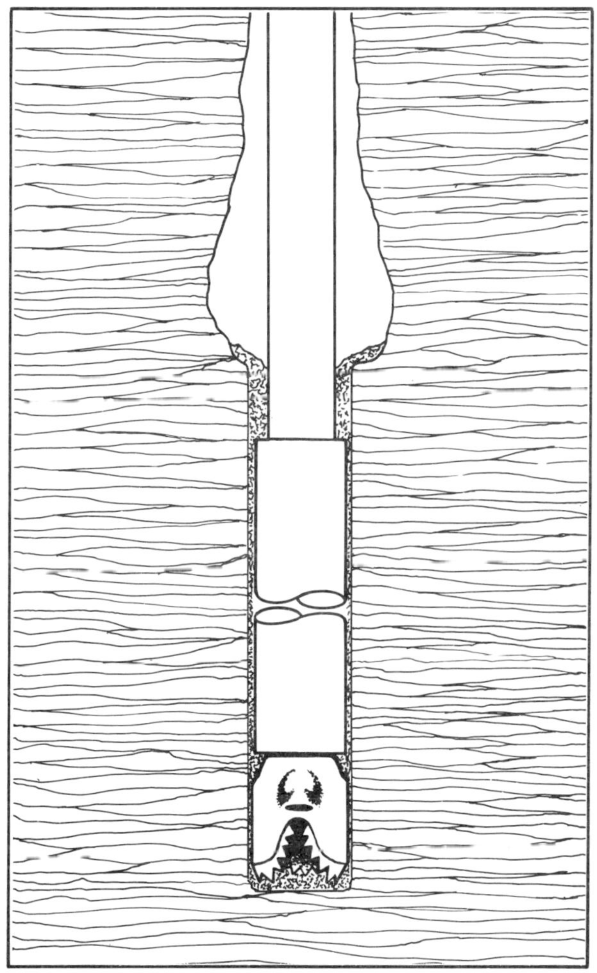

Figure 79. Pipe stuck in sloughing hole

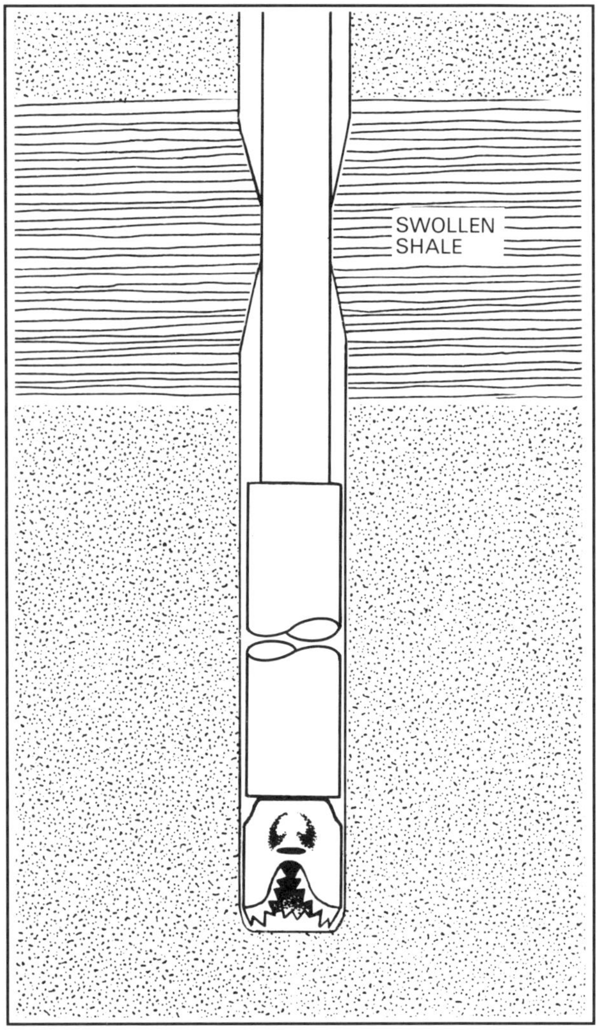

Figure 80. Stuck pipe caused by swollen shale

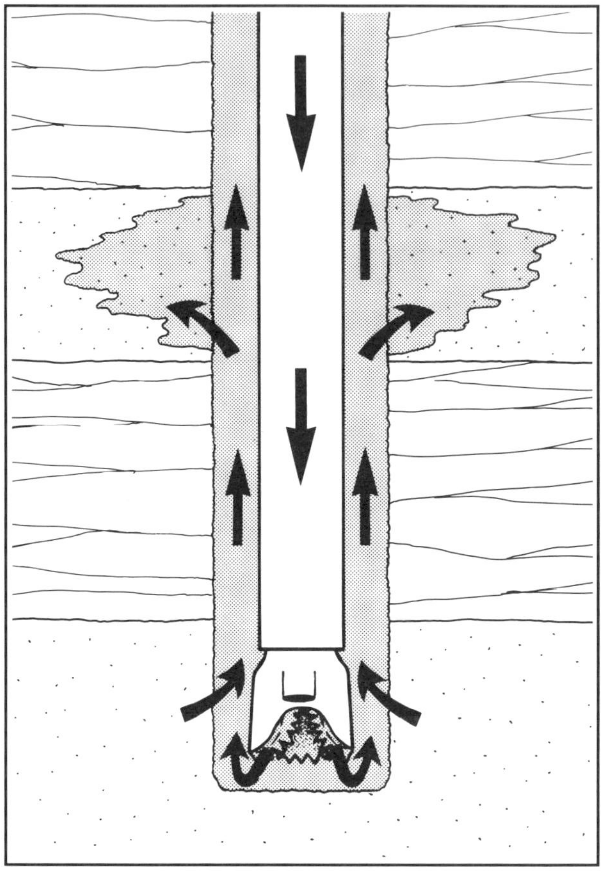

Figure 81. Lost circulation

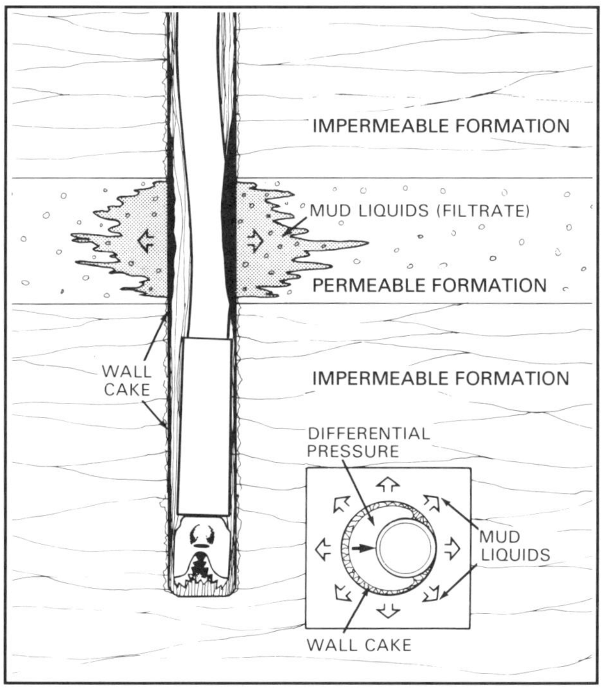

Figure 82. Differential sticking caused by slow loss of drilling fluid

Lost Circulation

When the hole has been drilled through a highly permeable formation, such as a porous sandstone or a cavernous limestone, there is a risk of losing drilling fluid to the formation (fig. 81). If drilling fluid is lost faster than it can be replaced, not enough mud may remain in the hole to hold back formation pressure at the bottom. The result may be a blowout.

Even if leakage into the formation is slow enough to control, the loss of liquid may cause mud solids to build up excessively on the hole wall, leading to *differential sticking* of the drill string (fig. 82). Like a sheet of paper stuck against a screen door, pressure can clamp the drill pipe against the wall of the hole and require an expensive recovery operation.

To control minor fluid loss to formation, the driller can add deflocculating, filtration control, or other lost circulation agents to the mud. Mud loss in significant or hazardous quantities may require anything from reducing mud weight to lower hydrostatic pressure (a risky procedure) to casing the hole through the lost circulation zone.

Pressure Control

Drilling unexpectedly into a high-pressure formation is one of the ways that blowouts happen. Seawater weighs about 9 pounds per gallon (ppg). It exerts a pressure of 4.65 psi at a depth of 10 feet, 46.5 psi at 100 feet, and 465 psi at 1,000 feet. The normal pressure gradient in formation is that of seawater—that is, pressure increases about 0.465 psi per foot of depth.

A hole full of 10-ppg drilling fluid will slightly overbalance normal formation pressure at all depths (fig. 83). At 10,000 feet, formation pressure will be 4,650 psi; mud hydrostatic pressure, 5,200 psi. Most wells are drilled slightly overbalanced to provide a safety margin against loss of pressure control.

Sometimes the drill bit enters a formation that is under greater-than-normal pressure—a porous formation isolated by impermeable rock and compressed by the weight of the overburden, a reservoir formation that outcrops at a higher elevation than the rig, or a highly compressed gas cap in a petroleum reservoir. When this happens, standard-weight mud in the well cannot keep formation fluid from entering the wellbore (fig. 84). When oil and gas contaminate water-based drilling fluid in the wellbore, they reduce its density and its hydrostatic pressure, allowing even more oil and gas to enter the wellbore.

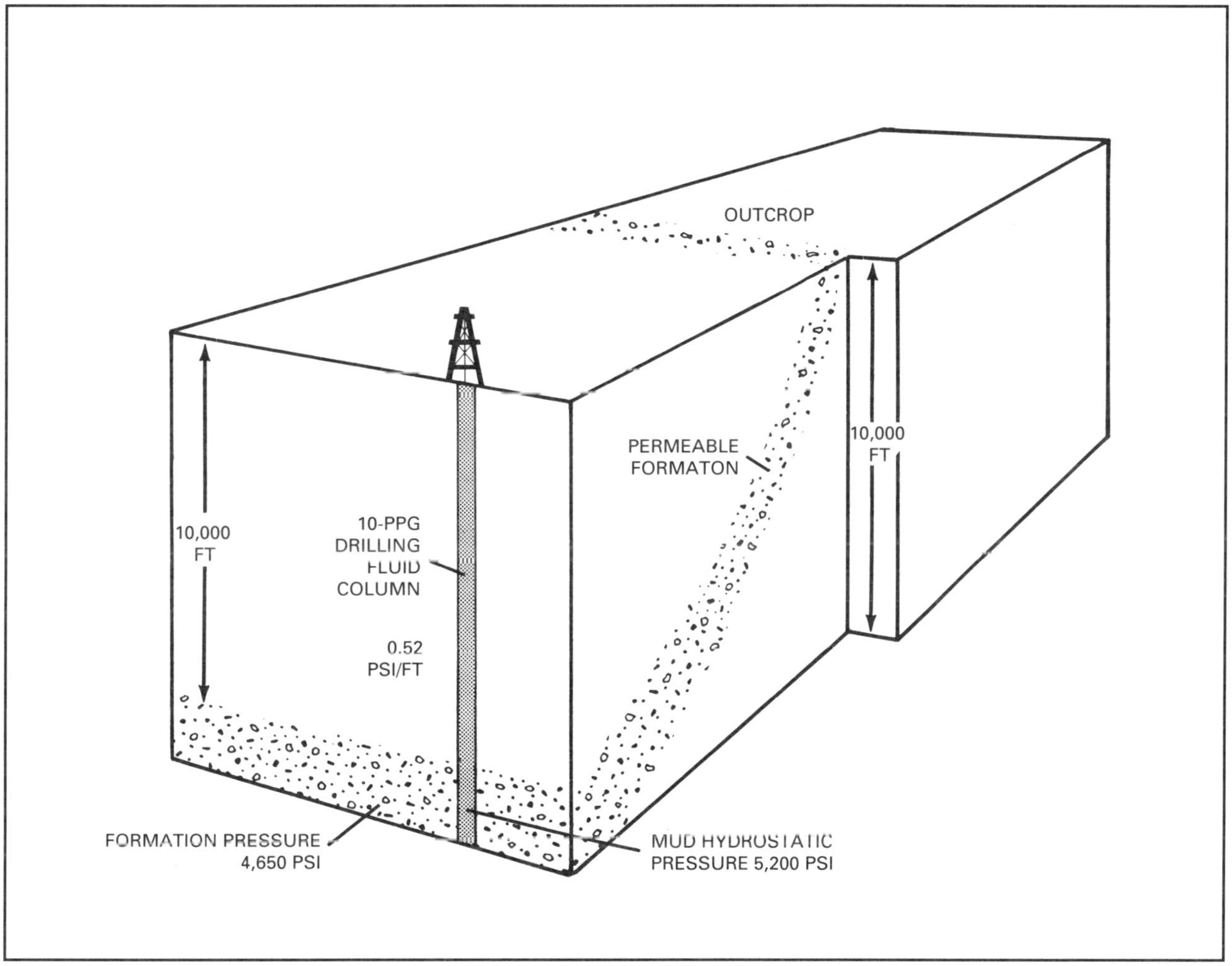

Figure 83. Hydrostatic pressures in a wellbore and in a formation

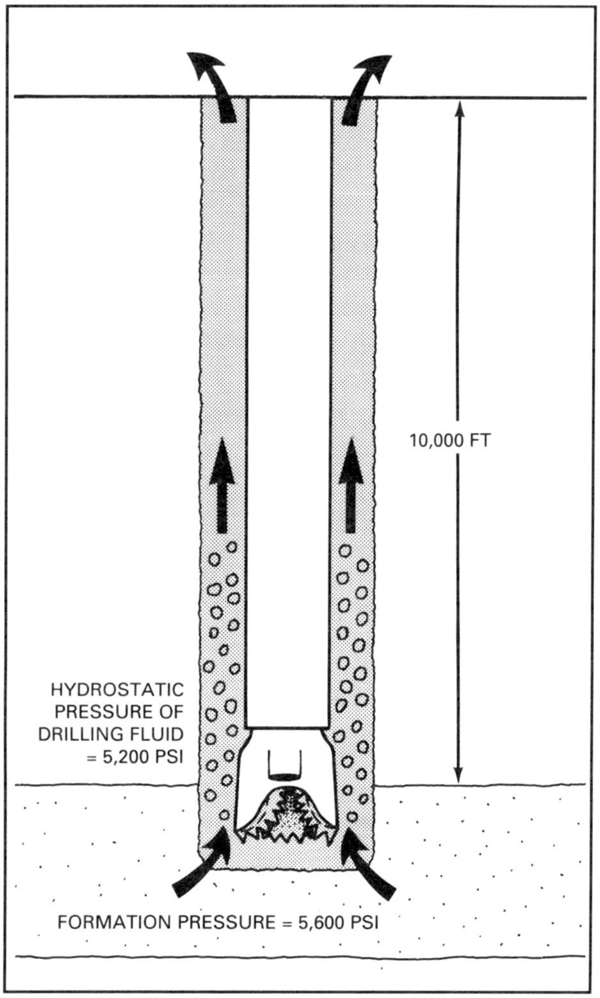

Figure 84. Formation fluid entering wellbore from a high-pressure formation

Unless the well is brought under control quickly, it will blow out.

Before the pressure imbalance becomes too great to control, the well must be shut in. The pumps are shut down, blowout preventers engaged, and choke lines closed to prevent further loss of drilling fluid. Then the driller begins to pump out the intruding formation fluid (the *kick*) under pressure, at the same time pumping in heavier drilling fluid to balance formation pressure. After the kick is out and the well filled with mud of sufficient weight, drilling can proceed normally with the blowout preventers and chokes open.

In some cases, as in shallow gas pockets of no economic significance, an abnormally pressured formation can be vented to the atmosphere at a controlled rate. Usually, though, blowing off gas pressure wastes both the marketable natural gas and the pressure that drives oil to the surface. Pumping oil is expensive, and it is impossible to pump all the oil out of a reservoir. But depleting reservoir pressure prematurely leaves oil in the reservoir that might otherwise be economically recoverable.

Most blowouts occur not in encounters with abnormal pressure but in routine drilling of normally pressured formations. All that is necessary for a blowout to happen is for well hydrostatic pressure to underbalance formation pressure. If mud is lost to a permeable formation, if too little drilling fluid is added to the hole when pulling out the drill string, if pressure is lowered by swabbing when tripping out—any of these situations may be enough to allow formation fluids under normal pressure to enter the hole. The more porous and permeable a formation, the faster a kick can develop and the more danger there is of a blowout.

Optimization

In the long run, the driller's job is to drill the well to the target as accurately, safely, quickly, and cheaply as possible. The continual adjustment of drilling factors to accomplish this goal is called *optimization*.

The six basic factors that affect the rate of penetration (ROP) are bit type, weight on bit, rotary speed, drilling fluid properties, bit hydraulics, and formation properties. The driller cannot control formation properties; however, the other five can be adjusted to achieve an optimum penetration rate. Therefore, knowing as much as possible about the formations to be drilled helps the driller decide how best to adjust the other factors.

Bit selection is one of the most important factors in optimization. A long-toothed offset roller cone bit, which makes hole by gouging out the rock, is best in soft formations, while a short-toothed, or insert, bit, which crushes and chips the rock, is better suited for medium to hard formations. A diamond bit, though expensive, often saves money by making hole faster in hard rock and by lasting longer, thereby requiring fewer bit changes. A fast-turning diamond bit on a downhole motor can drill hard formation quickly while minimizing wear on the drill string.

Weight on bit and rotary speed are usually considered together: when one is increased, the other is decreased (table 3). The optimum power that can be delivered by the bit is relatively constant and can be achieved either by high rotary speed and low weight on bit or by slower rotation with higher weight. Increasing or decreasing both at the same time causes either excessive bit wear and equipment damage or a slow penetration rate. For a 9½-inch insert bit in shale, the optimum combination may be a higher rotary speed (70 rpm) with low weight on bit (30,000 lb) or slower speed (55 rpm) with higher weight (40,000 lb).

The best drilling fluid is the lightest possible, with the lowest solids content. Sometimes, when the risk of a blowout is small, a well is drilled *underbalanced*—with underweight mud or clear water—so that the rock's own pore pressure helps break it apart. Drilling with pressurized gas or air, in locations where there is minimal danger of a blowout, is the fastest method. Fluid with high solids content tends to slow the rate at which rock fragments are detached from the bottom of the hole. High viscosity has a similar effect, but high gel strength helps clean cuttings from the hole. Adding oil to water-base mud helps reduce drill string friction and prevent sticking.

The hydraulic horsepower of the drilling fluid as it jets from the bit nozzles is vital to rotary bit performance (fig. 85). Increasing the cutting power of the bit is useless unless there is sufficient hydraulic power to remove the cuttings and carry them to the surface. Increasing hydraulic horsepower allows increases in both optimum weight on bit and rotary speed. Maximum benefit can be derived from a given combination of pump pressure and volume by installing the combination of nozzles that causes the maximum pressure drop from inside the drill pipe to the annulus at the bit. The greater the pressure drop, the higher

Table 3
Recommended Weight/Speed Combinations for Insert Bits

Formation Type	Bit Diam. (in)	High Speed/Low Wt. Rpm/Wt.	(1,000 lb)	Low Speed/High Wt. Rpm/Wt.	(1,000 lb)
Soft (unconsolidated sand)	6¼–6½	55	20–21	40	25–26
	7⅞	55	31	40	35
	8⅜–8¾	55	34–35	40	39–40
	9½–9⅞	55	38–40	40	44–45
	11	55	43	40	50
	12¼	55	47	40	54
Medium soft (limestone)	6¼–6½	65	20–21	45	27–28
	7⅞	65	31	45	39
	8⅜–8¾	65	34–35	45	42–44
	9½–9⅞	65	38–40	45	48–49
	11	65	43	45	50
	12¼	65	47	45	53
Medium (shale)	6¼–6½	70	15–16	55	20–21
	7⅞	70	23	55	32
	8⅜–8¾	70	27–28	55	34–36
	9½–9⅞	70	30–32	55	39–40
	11	70	33	55	44
	12¼	70	34	55	44

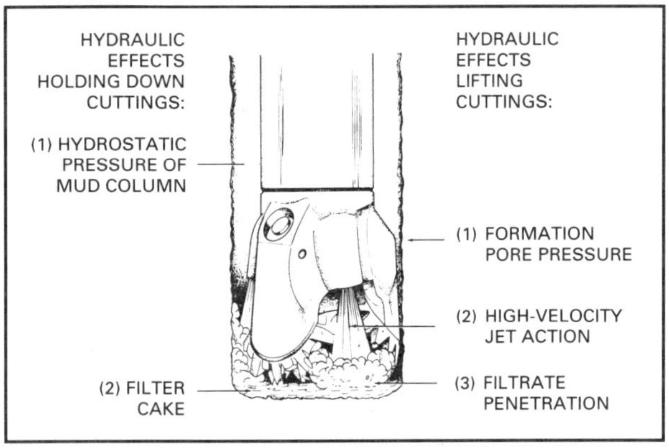

Figure 85. Effects of hydraulics on hole bottom

the velocity of the jet that carries away the cuttings. The main limitation is that enough pressure must remain to circulate the drilling fluid and its load of cuttings from the bottom of the hole to the surface.

Completions

Once a well has been drilled to the target depth and the reservoir is found to have commercial potential, geologic factors figure into the decision on how the well is to be *completed*, or made ready for production. The amount and type of hydrocarbons present, natural drive strength, and reservoir rock porosity and permeability are some of the factors used in determining whether it is economically feasible to complete a well. Other factors, such as the integrity of the formation, the number and vertical distribution of producing zones, the size of sedimentary grains, and the physical and chemical properties of the formation fluids, influence the selection of tubular goods and other equipment that will be installed in the producing interval to maintain production and prevent damage to the reservoir and the well.

The most common method of completing a well is to run casing or liner in the producing horizon and, using either projectiles or shaped explosive charges, perforate the pipe and the nearby formation in several places (fig. 86). The perforated pipe holds the formation in place while fluids flow into the well through the perforations. In a sandstone reservoir formation, where loose sand might clog the bottom of the hole and destroy production tubing and valves, a wire-wrapped screen can be used, or the completion can include a surrounding gravel pack grading from fine gravel near the liner to coarse gravel at the outer perimeter. The *barefoot* (or *open hole*) *completion*, which has no liner or casing in the producing interval, is sometimes used to produce oil and gas from a limestone reservoir.

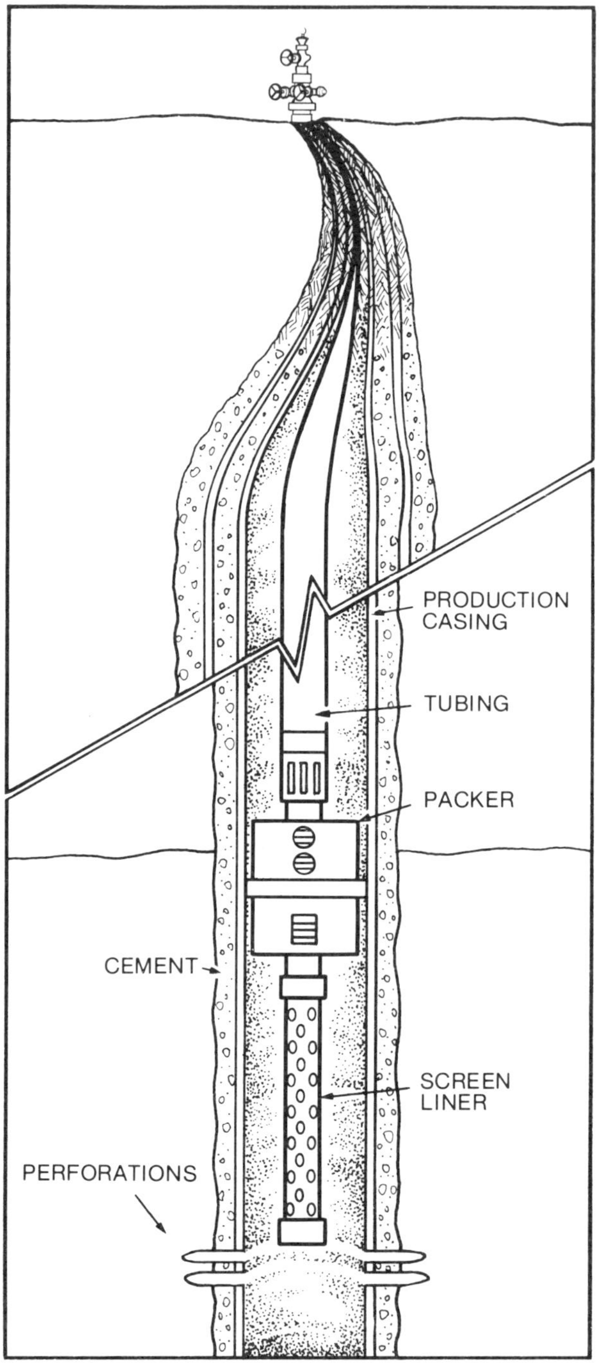

Figure 86. Conventional completion

REVIEW QUESTIONS
Chapter 6

______ 1. Of the following factors that affect the speed and efficiency of drilling, which *cannot* be changed by the driller?

- A. weight on bit
- B. rotary speed
- C. mud weight
- D. geology

______ 2. (T/F) Most blowouts occur when drilling into abnormally pressured formations.

______ 3. In a barefoot completion—

- A. the producing zone is not cased.
- B. a wire-wrapped screen is installed.
- C. the tubing is perforated.
- D. the driller does not wear shoes.

______ 4. When drilling through rock layers dipping more than 45°, the bit tends to—

- A. drill updip.
- B. drill perpendicular to the formation.
- C. drill parallel to the layers.
- D. drill to the left.

______ 5. The main advantage of drilling underbalanced is that—

- A. it provides a margin of safety against blowouts.
- B. it allows the rock's own pore pressure to break out chips.
- C. it enables the driller to apply more weight to the bit.
- D. it allows formation fluids to come to the surface.

6. The adjustment of drilling factors to make hole as efficiently as possible is called ______________.

______ 7. Which of the following types of rock is most likely to lead to lost circulation?

- A. dense shale
- B. cavernous limestone
- C. massive granite
- D. heavily cemented sandstone

______ 8. (T/F) The rotating bit has a natural tendency to "walk" to the left in the plane of the formation being drilled.

9. Normal formation pressure at 5,000 feet is __________ psi.

______ 10. (T/F) Blowing off gas pressure from an oil reservoir decreases the amount of oil that can be recovered.

A 11. When drilling through sedimentary layers that are tilted less than 45°, the bit tends to drill—

(A) perpendicular to the layers.
B. parallel to the layers.
C. vertically, regardless of formation dip.
D. toward the south.

12. A diamond bit works best in hard formations.

B 13. When oil or gas enters the wellbore from the formation—

A. it makes the drilling fluid heavier.
(B) it pushes drilling fluid out of the wellbore.
C. it lubricates the bit and increases the penetration rate.
D. control of the well is immediately lost.

A 14. Drilling through horizontal layers of rock of different strength may result in—

(A) crooked hole.
B. a blowout.
C. a broken bit.
D. differential sticking.

C 15. When shale absorbs water, it often—

A. shrinks.
B. hardens.
(C) swells.
D. lithifies.

CHAPTER 7
PRODUCING THE RESERVOIR

Introduction

A familiarity with geology is essential in getting the maximum return on a drilling investment, from the first regional gravity anomaly to the last drop of oil squeezed from a reservoir. Geology does not cease to be a factor simply because the reservoir has been located. The ultimate value of a discovery depends on the techniques used to produce and maintain the well and the measures taken to extend the productive life of the reservoir; the techniques chosen depend largely on the area's geological character.

To be commercially producible, a reservoir must meet three basic requirements. First, the reservoir rock must be porous and permeable enough to allow the fluids to flow readily to the wellbore. Second, the proportion of oil or gas among the formation fluids must be great enough to ensure that the well is profitable. Third, formation pressure, from either water or gas, must be sufficient to force oil or gas into the wellbore—that is, there must be a drive.

Porosity and Permeability

Porosity is expressed as the percentage of void space in total reservoir rock volume. Total porosity, however, is not as important as *effective porosity,* the percentage of rock volume represented by interconnected pore space. Petroleum can accumulate in, and be produced from, only those pores that are connected. Isolated pores simply retain connate water. Total porosity can vary from less than 5% in a tightly cemented sandstone or carbonate formation to more than 30% in a well-sorted or unconsolidated sand. In general, reservoir rocks should have at least 10% effective porosity to be considered commercially producible.

The permeability of reservoir rock is the ease with which fluids can move through its interconnected pore spaces (fig. 52). If the connections (pore throats) are narrow and formation fluids move with difficulty, the well may flow very slowly or require artificial lift. Well stimulation techniques may be employed—and in many instances are routinely applied as soon as the well is completed—to increase permeability and speed production. The basic unit of permeability is the *darcy.* Most petroleum reservoirs have permeabilities measured in thousandths of a darcy, or *millidarcys* (md). Although high porosity is often accompanied by high permeability, the correlation is far from exact.

Reservoir Fluids

Oil, water, and gas are *fluids*—substances that flow at ordinary temperatures and pressures. Oil and water are also *liquids*—incompressible fluids. Gas is a fluid but not a liquid.

Most porous rock at depth contains salt water that was trapped when seafloor sediments were lithified. In permeable rock containing petroleum, much of this original salt water has been displaced by migrating oil, gas, and water. Thus, although the salt water in a reservoir is often called "connate" water, it is more properly referred to as "interstitial" water.

Oil droplets in water-saturated reservoir rock tend to join together in larger, more buoyant droplets. As they rise and coalesce in the higher parts of the reservoir, they displace interstitial water downward.

Oil accumulations usually contain gas in solution. The lower the pressure and the higher the temperature, the less gas remains dissolved in oil. If the amount of dissolved gas reaches saturation, free gas accumulates at the top of the reservoir and displaces the oil downward.

There are three basic types of hydrocarbon reservoirs: oil only, oil and gas, and gas only (*nonassociated* gas) (fig. 87). Since gas is the least dense of the fluids, any free gas present is always at the top of the pool in the form of a *gas cap.* Beneath the gas is the oil zone, which is underlain by water.

Water is usually present in both the oil and the gas zones of a reservoir. Because of its wetting power, water clings more tenaciously to the grains of rock

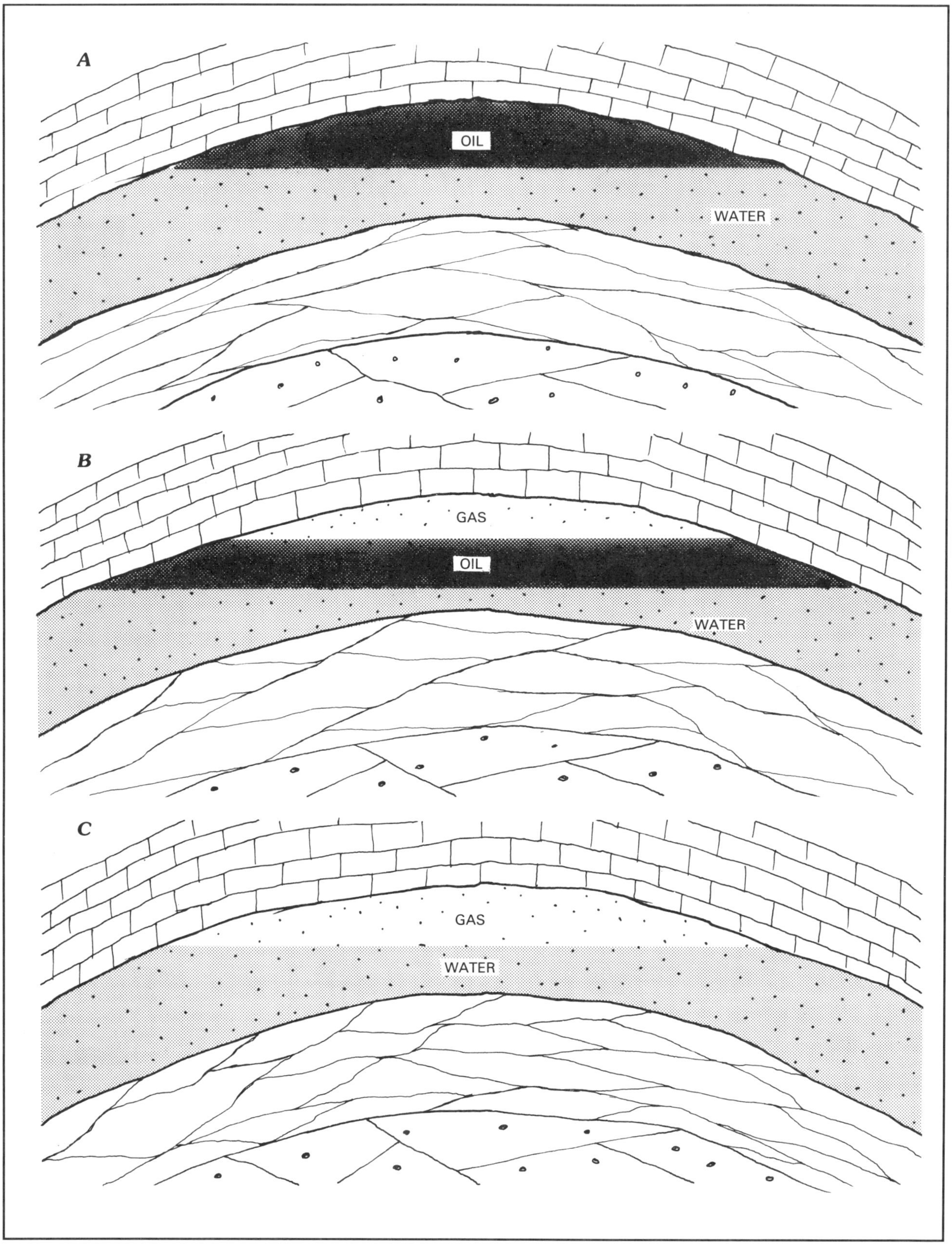

Figure 87. Types of hydrocarbon reservoirs: (*A*) oil only; (*B*) oil and gas; (*C*) nonassociated gas

than do the other fluids. With the exception of a few "oil-wet" reservoirs, petroleum reservoirs are all "water-wet," with water coating the grains throughout. The smaller the pores and pore throats, the greater the percentage of pore space occupied by this water film. Water that entirely fills the smaller pore throats can cut reservoir rock's effective permeability for oil to zero. Thus, even permeable rock, when water-saturated, can trap oil.

The contact between the oil and the water zones in a reservoir is not a distinct boundary but a zone in which the proportions of fluids change gradually (fig. 88). Upward from the top of the water zone, the proportion of oil rises from near zero to a maximum of perhaps 70% over a vertical range of 10 to 15 feet. Continuing upward, the mixture in the water-wet reservoir stays close to 70% oil and 30% water until the bottom of the oil-gas contact is reached. Here begins a much thinner zone, only 2 or 3 feet thick (because of the great difference in density between oil and gas), in which the oil falls to a small percentage while the gas reaches 70% or more of the total fluid volume.

Within a reservoir formation, fluids flow from one point to another because of gravity, capillarity, and pressure. The force that causes oil and gas to migrate upward is gravity. Although this migration seems contrary to the downward pull of gravity, it is simply a reflection of differences in density. The heaviest fluid, water, responds more strongly to gravity and displaces lighter fluids upward.

A *pressure gradient* is a uniform, continuous change in pressure from one point to another. Fluids in the high-pressure zone are driven toward lower pressure. When a well is drilled into a fluid-bearing

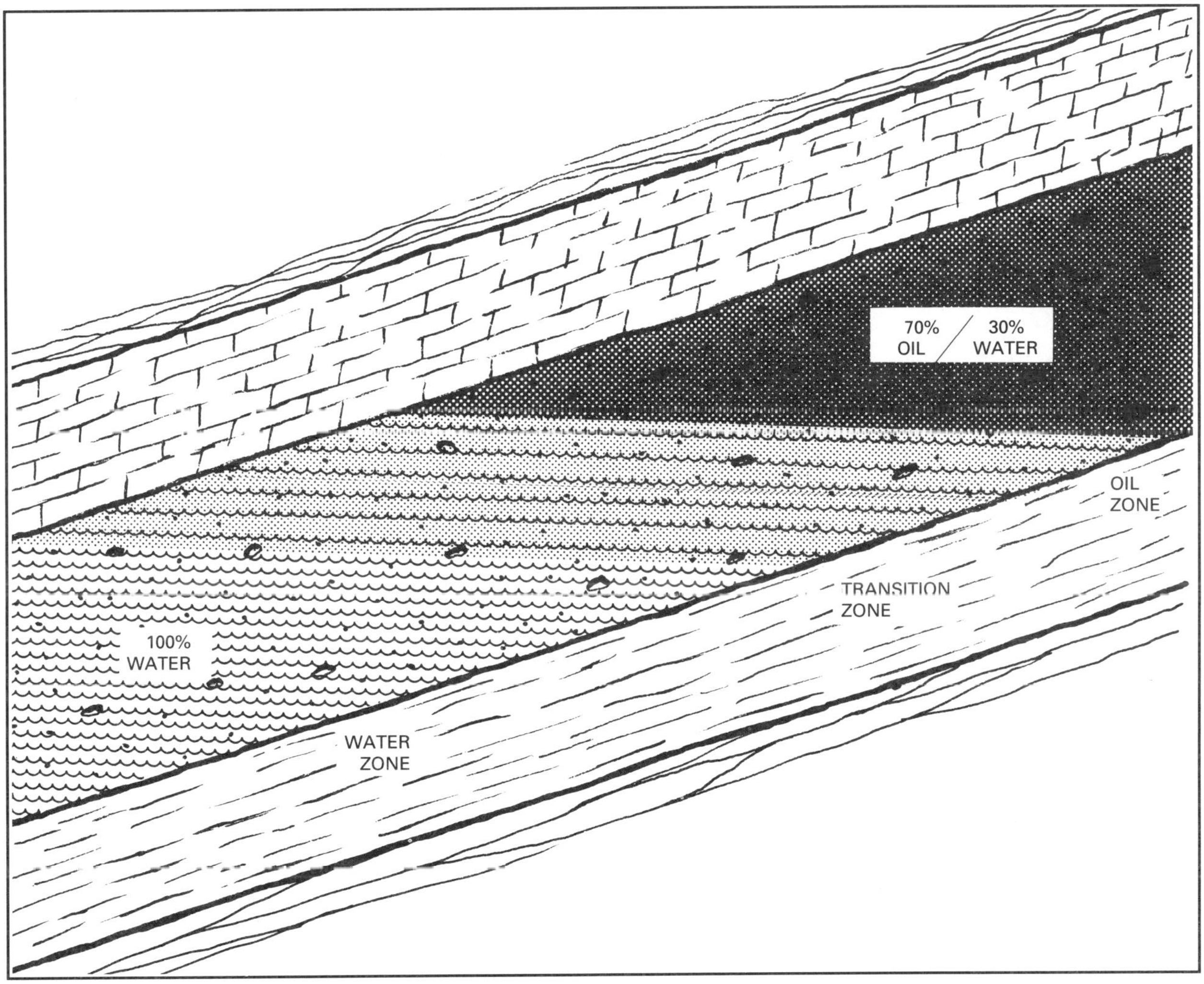

Figure 88. Oil-water contact, or transition, zone

formation and the pressure in the wellbore is lower than formation pressure, formation fluids flow toward the wellbore.

Capillarity is the physical attraction between mineral grains and water that pulls water into small pore spaces and pore throats. This force is sometimes called *capillary pressure*; it is actually a multitude of microscopic pressure gradients that drive oil out of the smallest pore spaces.

Drives

Reservoir drive is the force that causes petroleum to flow out of the reservoir rock and into the wellbore. Most petroleum reservoirs flow unassisted for at least the first part of the reservoir's productive life. If this natural drive is used up, production can continue only by artificial lift or by reservoir flooding, either of which may be prohibitively expensive.

There are two basic types of drive: *depletion drive*, which operates until pressure is no longer sufficient to drive oil out of the well; and *water drive*, in which the growing contamination of the wellstream by water eventually overcomes the relatively constant drive pressure. Most reservoir drives are a combination of these types.

A pure depletion drive operates in a closed reservoir, where the hydrocarbons are not in contact with a large body of permeable water-bearing sand, or where the reservoir rock has limited permeability. The energy that drives the oil into the wellbore comes from the expansion of gas (fig. 89). When the pressure falls, as happens when a well is drilled into the reservoir, gas dissolved in the oil begins to come out of solution, forming tiny bubbles. These bubbles expand, displacing the oil, which flows into the wellbore along with the entrained gas. This mechanism is termed a *solution-gas drive*.

At first, the hydrocarbon accumulation in a solution-gas drive reservoir is composed entirely of oil. Any gas present is not apparent because it is dissolved in the oil, and reservoir pressure is too high for gas bubbles to form. When the pressure is sufficiently reduced, however, the dissolved gas reaches the

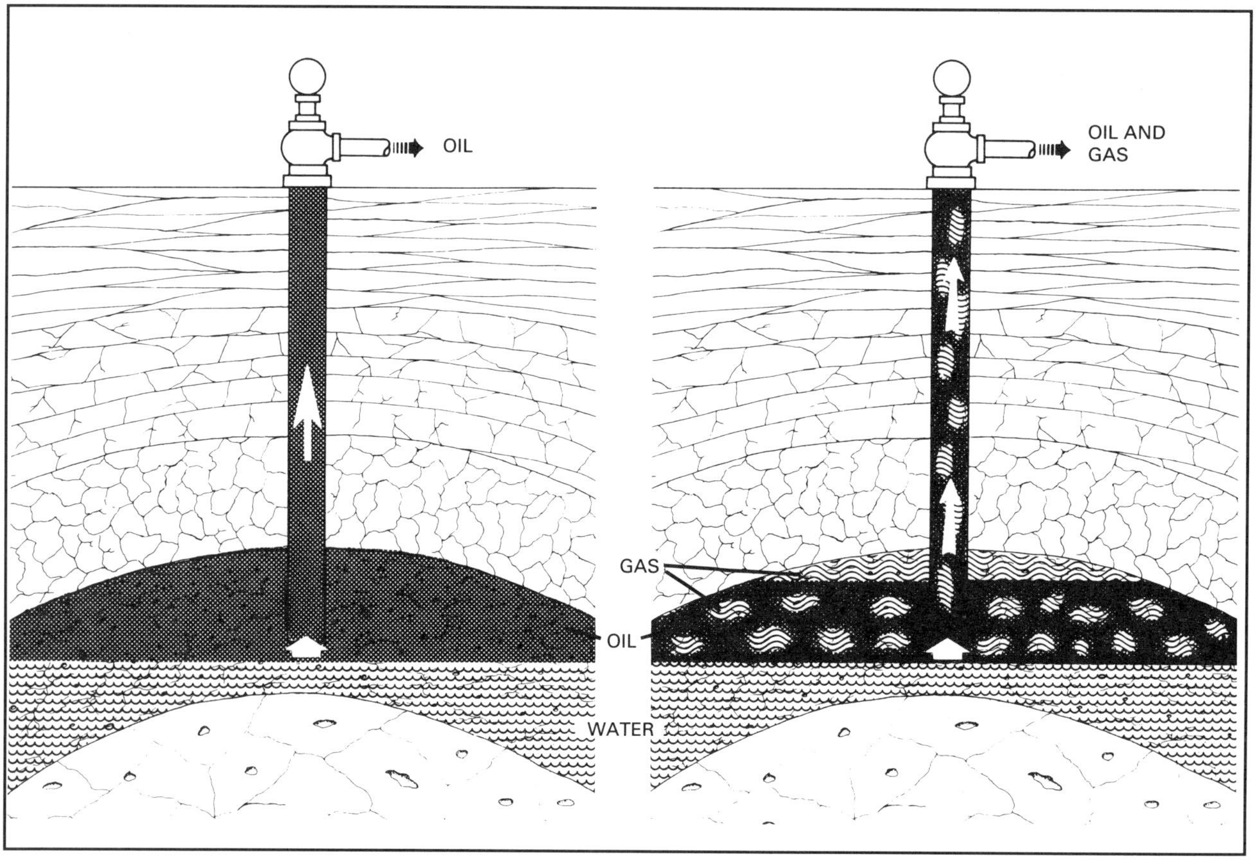

Figure 89. Solution-gas drive reservoir

saturation point and bubbles begin to form. As these bubbles expand, they push the oil out of the reservoir and up to the surface.

Early in the production history of a typical solution-gas well, each barrel of oil that reaches the surface may be accompanied by 100 cubic feet of gas. This gas-oil ratio may rise to several times that value as pressure drops, until finally both oil and gas production fall to noneconomic values (fig. 90). In a typical pure solution-gas drive reservoir, (1) reservoir pressure declines rapidly and continuously; (2) the gas-oil ratio begins low, rises to a maximum, and drops; (3) little or no water is produced; (4) pumping or artificial lift is required early; and (5) recovery of the original oil in place ranges from 5% to 30%.

A *gas-cap* drive, another type of depletion drive, operates by the expansion of the gas cap above a gas-saturated oil accumulation (fig. 91). Natural pressure is maintained longer than in a solution-gas drive, so dissolved gas comes out of solution more slowly as the oil is produced. As the gas cap expands, it forces the gas-oil contact downward until the gas cap itself begins emptying into the wellbore. To delay this encroachment as long as possible, wells in a gas-cap drive reservoir are completed as deep in the oil zone as possible.

Oil production rates in a gas-cap reservoir do not decline as sharply as in solution-gas reservoirs, because there is more gas available (fig. 92). Gas-oil ratios also rise less sharply, at least until the gas-oil contact reaches the producing zone. In a typical gas-cap drive reservoir, (1) reservoir pressures fall slowly and continuously; (2) surface gas-oil ratios rise slowly and continuously; (3) little or no water is produced; (4) depending on the size of the gas cap, flow life may be long; and (5) recovery of original oil in place ranges from 20% to 40%.

After a time, the gas pressure in a depletion drive falls too low to drive the oil up the wellbore to the surface. Some of the remaining oil can usually be recovered by artificial lift, but at some point the energy required to pump the oil to the surface may exceed the market value of the oil. Early drillers who penetrated the gas zone of a reservoir would blow off

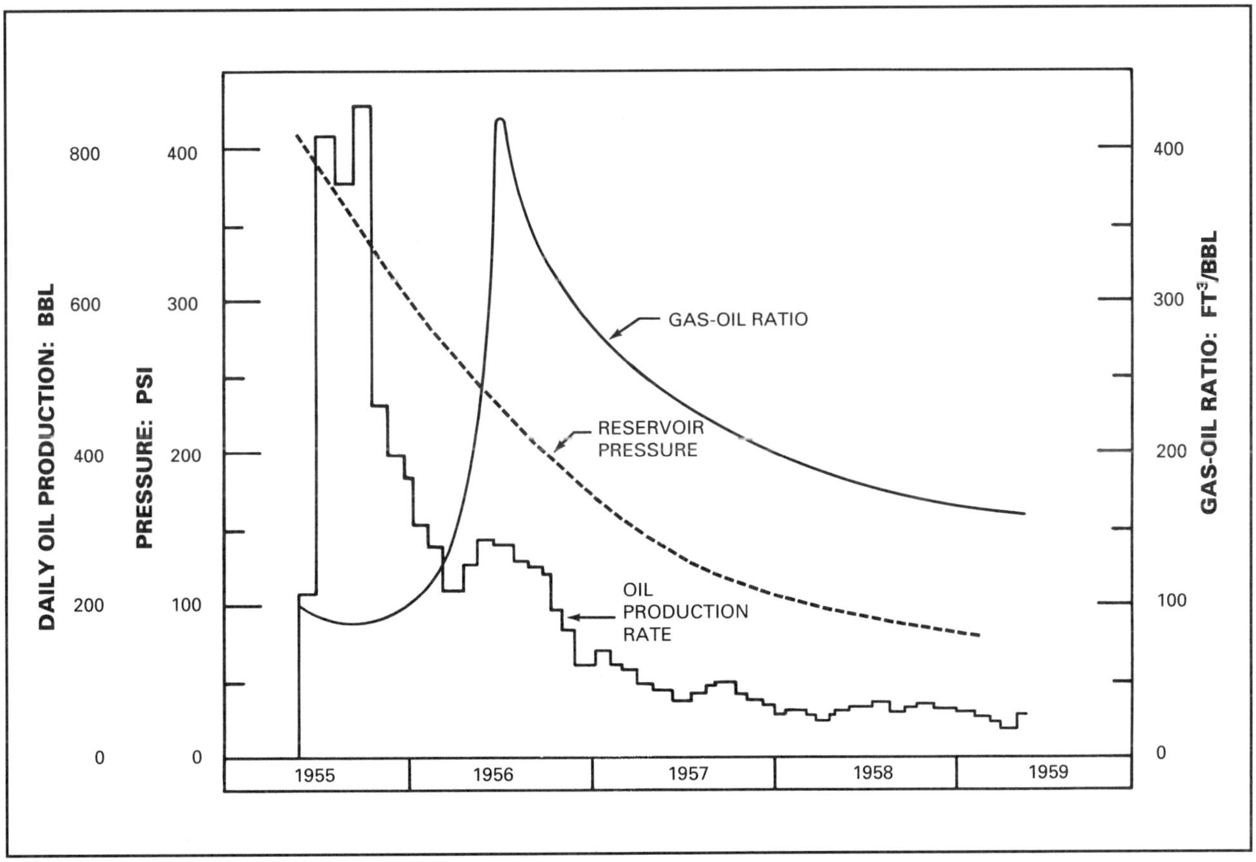

Figure 90. Production data from a solution-gas reservoir

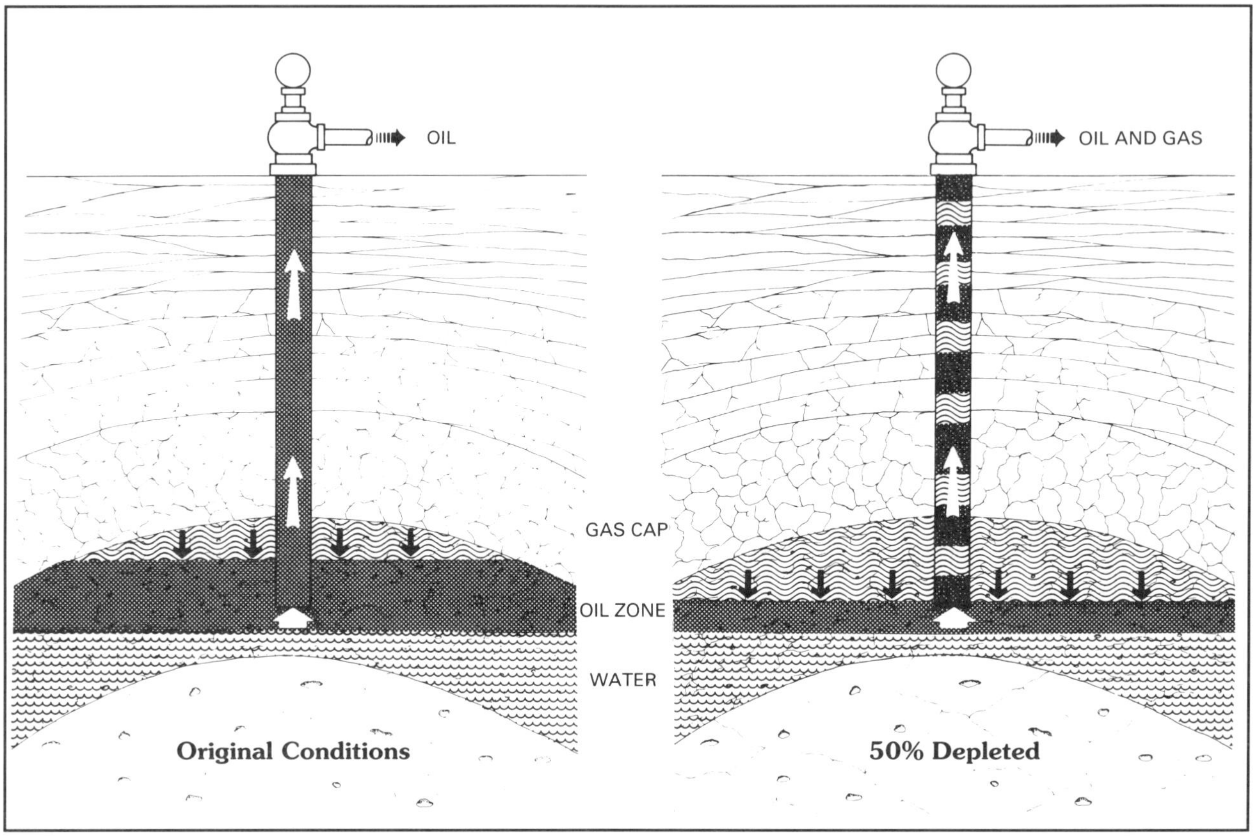

Figure 91. Gas-cap drive reservoir

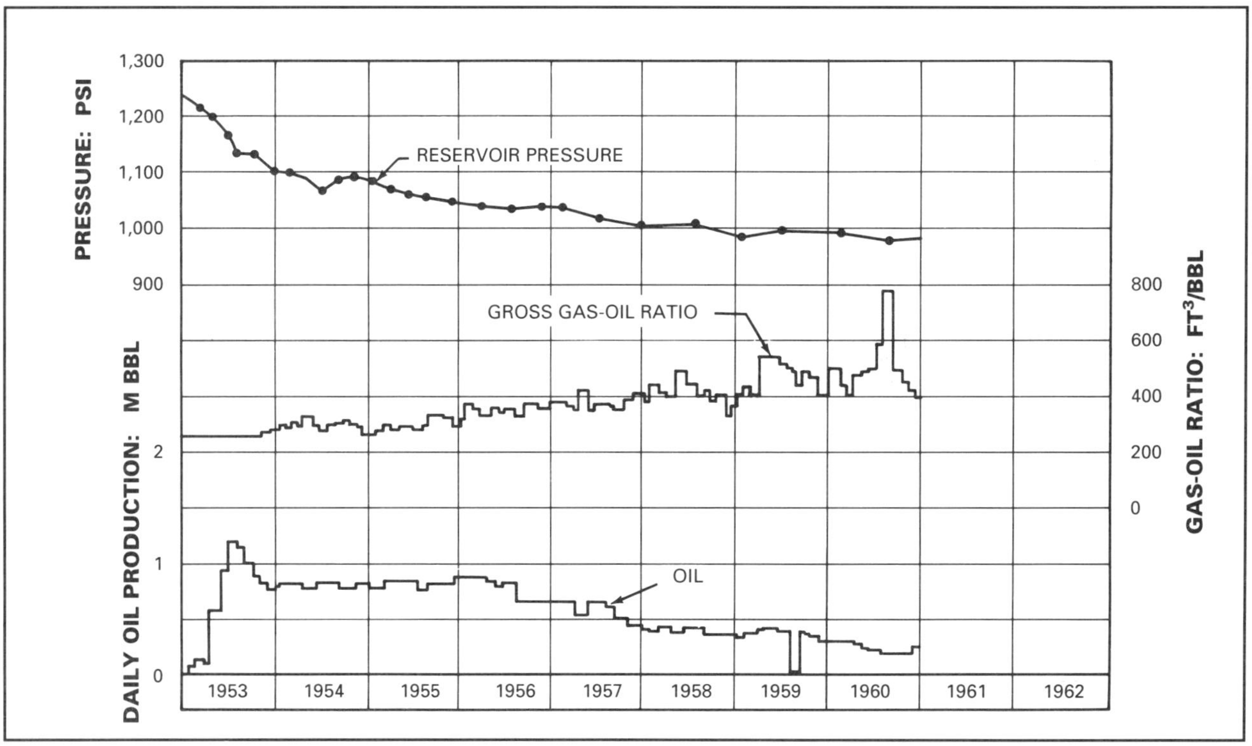

Figure 92. Production data from a gas-cap reservoir

the gas cap so that oil would rise into the gas zone. As a result, huge reserves of oil were lost because they could no longer be produced economically.

Water drive is more efficient than depletion drives in terms of producing more of the oil in place (fig. 93). Most petroleum reservoirs have water either beneath or downdip from the oil zone. Water occurring beneath the oil zone is *bottom water*; along the edge of the oil zone, on the flanks of the structure, it is called *edgewater*.

A reduction of pressure in a water drive reservoir allows bottom- or edgewater to migrate upward, pushing oil ahead of it into the wellbore. The pressure exerted by the water is usually normal pressure, the equivalent of a column of seawater extending from the surface to reservoir depth—0.465 psi/ft.

Under ideal conditions of pressure, oil viscosity, and reservoir porosity and permeability, water is very efficient at displacing oil from the reservoir (fig. 94). Drive pressure stays nearly constant, as does the surface gas-oil ratio, because little or no gas comes out of solution in the reservoir. The ratio of water to oil slowly increases, reducing the oil production rate (unless field development includes facilities to handle increasing amounts of water).

As water replaces the oil that is withdrawn, it begins to encroach on the wells in the lowest part of the producing zone. The faster oil is produced, the more likely water is to rise into the part of the reservoir near the well. This phenomenon is called *coning* (fig. 95). The amount of water produced with the oil rises until the hydrostatic pressure of the water in the well stream overcomes the natural drive pressure. This kills the well. Production from wells higher in the structure is not affected. Ideally, the last producing well would be the one in the highest part of the structure, which would be flooded out as the last of the oil is produced; however, the ideal is seldom approached. The permeability of reservoir rock varies; water encroaches more rapidly in the most permeable zones, bypassing pockets of oil instead of driving them into the well (fig. 96).

In a typical edgewater drive reservoir, the zone of oil accumulation is relatively thin, with oil occupying the entire thickness of the formation and water occurring only along the flanks of the oil zone

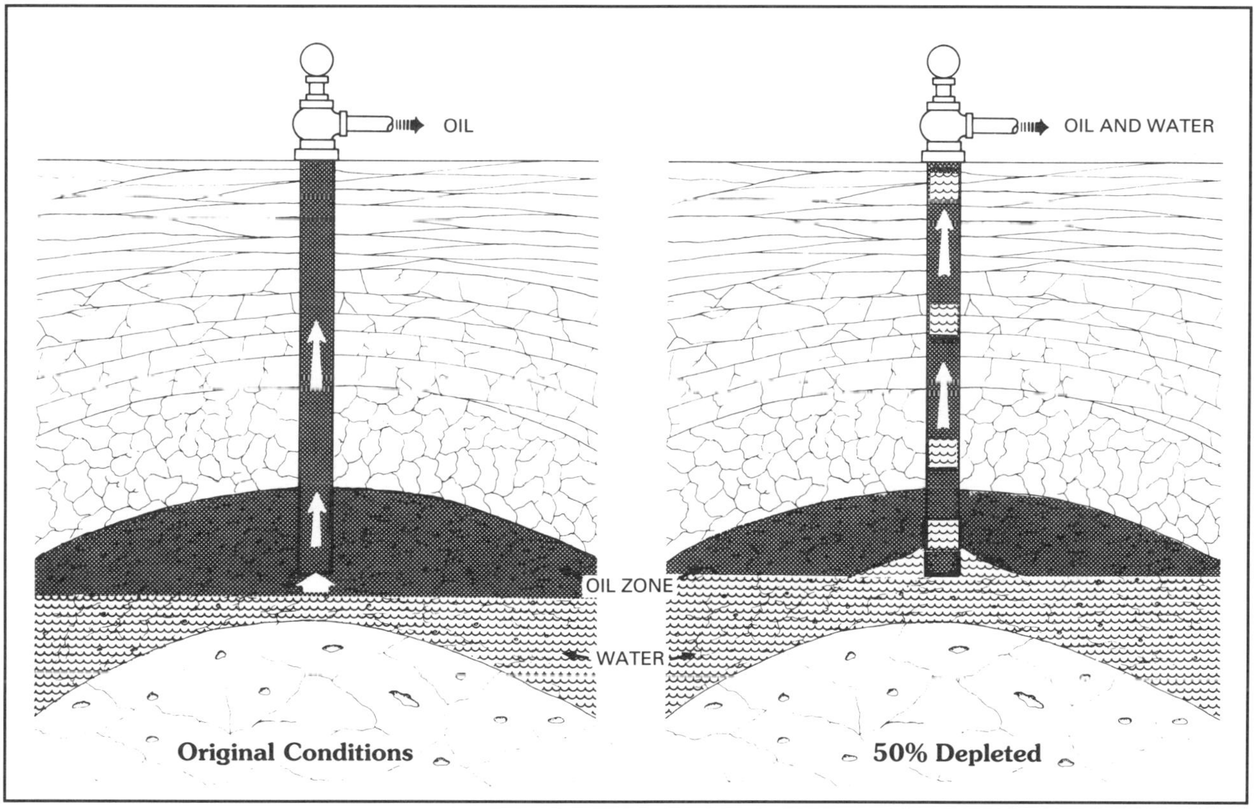

Figure 93. Water drive reservoir

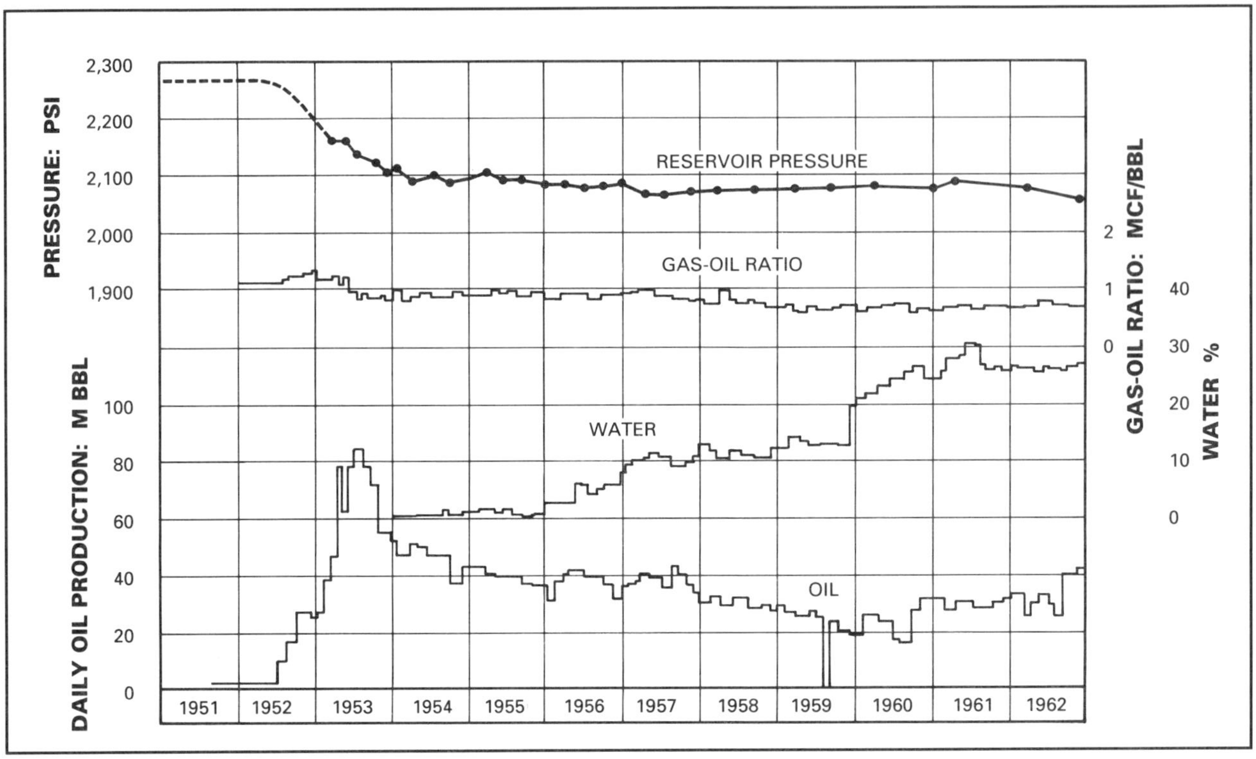

Figure 94. Production data from a water drive reservoir

(fig. 97). Wells drilled near the top of the structure will encounter only oil; wells along the flanks will penetrate water. In a bottom-water reservoir, the formation is thick enough, or the oil accumulation is thin enough, for water to underlie all of the oil zone (fig. 98). A well

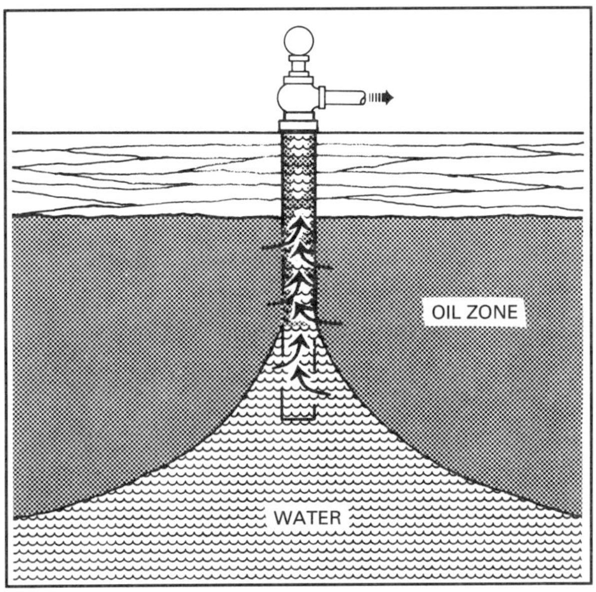

Figure 95. Coning of water into a well

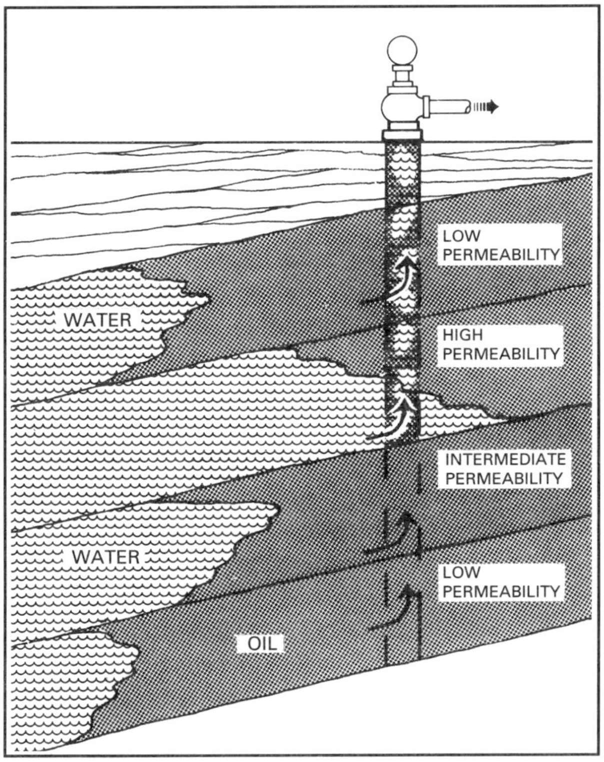

Figure 96. Irregular water encroachment and early breakthrough in rocks of high permeability

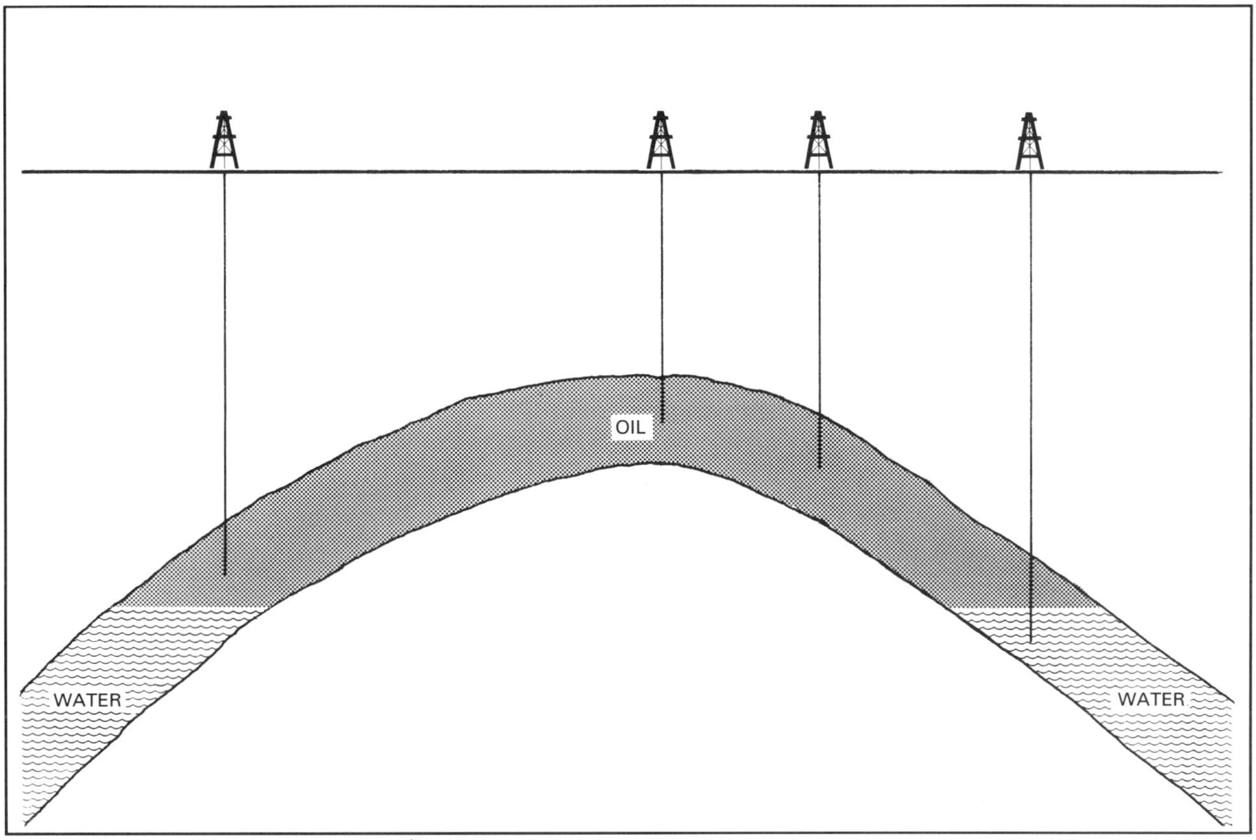

Figure 97. Edgewater drive reservoir

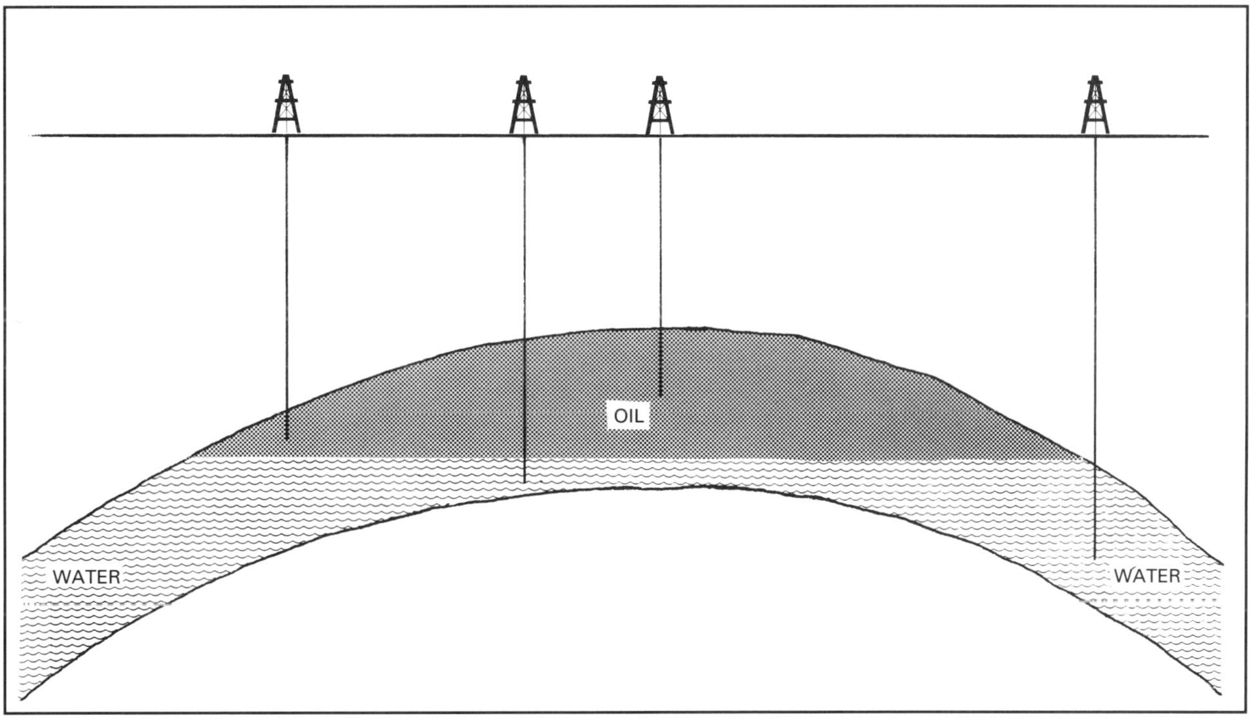

Figure 98. Bottom-water drive reservoir

drilled through the oil zone anywhere in a bottom-water reservoir will penetrate first oil, then water. In general, edgewater drive is more efficient at producing the oil in place, because it is difficult to keep water from coning and entering a well in a bottom-water reservoir.

In water drive reservoirs, (1) reservoir pressure remains high as long as the volume of oil withdrawn is replaced by a similar volume of water; (2) if reservoir pressure remains high, the surface gas-oil ratio remains low; (3) water production may start early and increase to an appreciable amount as water encroaches into the oil column; (4) wells tend to flow until water production becomes excessive and kills the well; and (5) up to 75% of the oil in the reservoir is produced.

Both water and gas are usually involved in producing oil from a reservoir. A typical *combination* drive has water below and a gas cap above the oil zone (fig. 99). If production is slow enough, water displaces the oil as it is produced and the gas cap does not expand. If production exceeds a certain rate, pressure drops, the gas cap expands, and the depletion mechanism comes into play.

Water drive reservoirs may be rate-sensitive (fig. 100). If the oil is withdrawn faster than water can replace it, pressure declines and gas begins to come out of solution. The water drive then becomes a combination drive, with solution gas providing part of the drive energy. If production rates increase, the reservoir may behave as a pure depletion drive reservoir; if production is slowed, pressure may recover and water drive become predominant again.

In a shallow, steeply dipping reservoir, a third drive mechanism, *gravity drainage,* may operate, especially late in production life. Oil, like any liquid, will run downhill under the influence of gravity. If gas pressure is depleted and water has not encroached, the oil remaining in the nearly empty reservoir may simply drain to the lowest point. This mechanism can be very efficient at producing the last drop of oil from a highly permeable depletion drive reservoir.

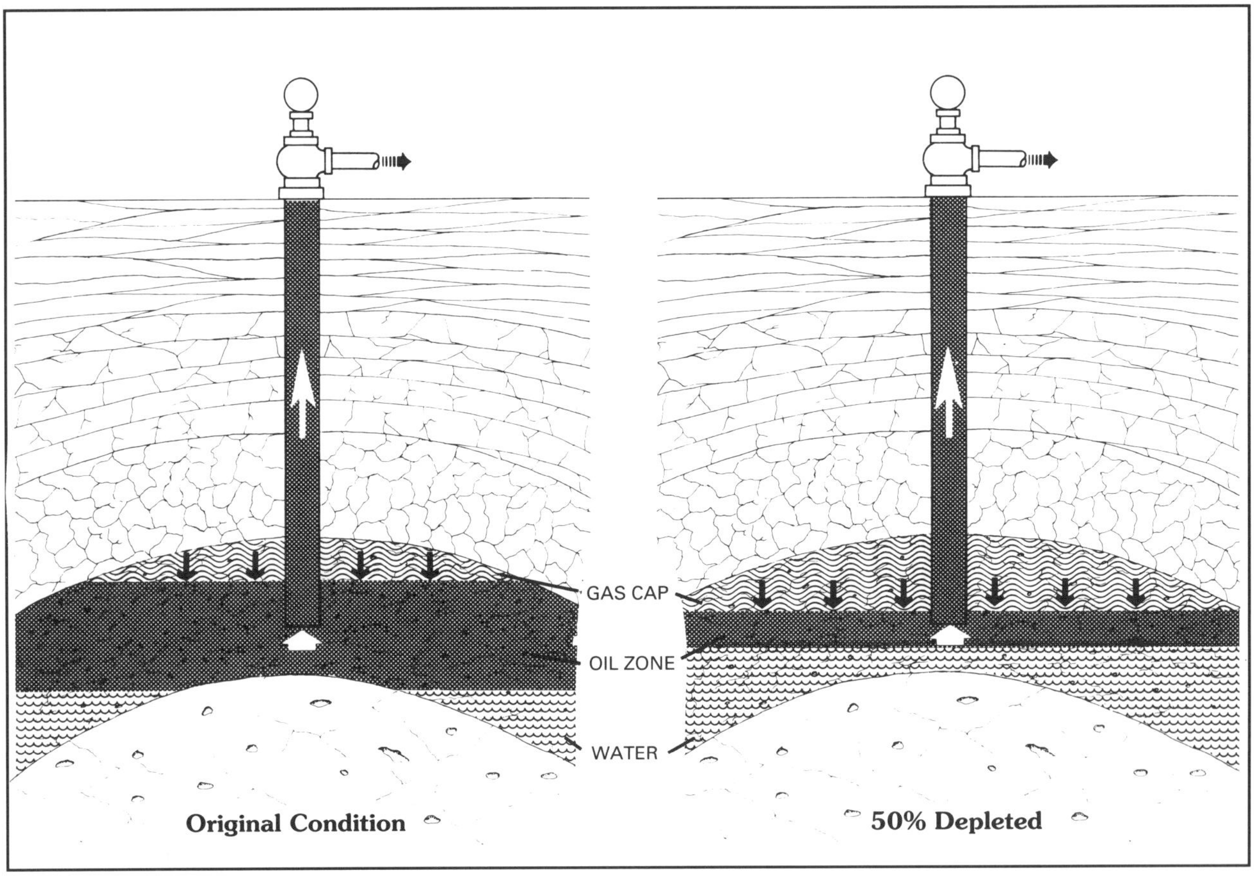

Figure 99. Combination water and gas-cap drive

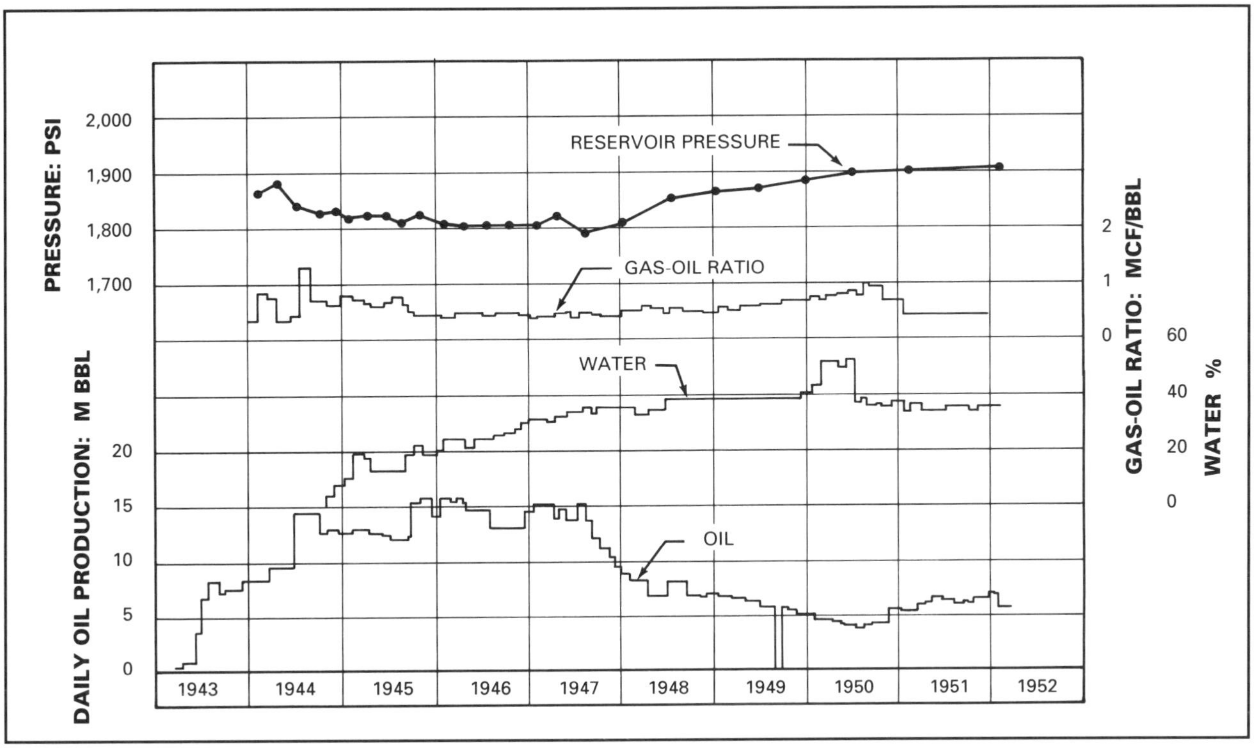

Figure 100. Production data from a water drive reservoir showing sensitivity to production rate

Characteristics of Producing Wells

A producing well is described in terms of three trends: pressure, gas-oil ratio, and water production. Each of the basic drive mechanisms has a characteristic pattern of trends. Part of the job of managing production for best recovery is anticipating these trends for each well or field and compensating for any problems that occur.

The pressure in a producing reservoir is lowest around the wellbore. In a typical well in reasonably porous formation, pressure may drop from 3,000 psi at some distance from the well to 1,900 psi inside the production casing. Most of this 1,100-psi pressure drop may occur within 15 feet of the wellbore. Fluids converging on the wellbore from a reservoir drainage area of 40 acres must flow faster as they funnel through the last few feet of the completion zone (fig. 101). In this type of flow (*radial*), velocity increases toward a *pressure sink*—the wellbore.

Because of the hydrostatic pressure of fluids in the tubing, pressure drops from the completion zone to the wellhead. In a 4,000-foot well with a bottomhole

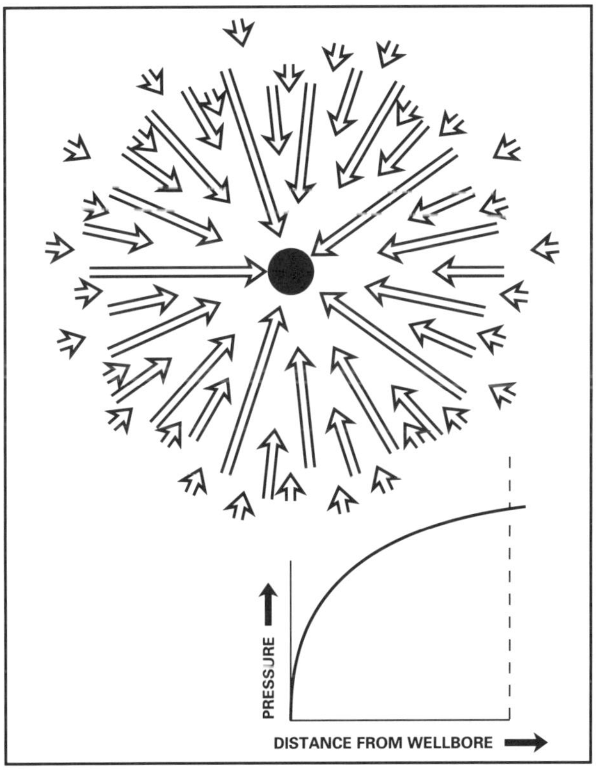

Figure 101. Radial flow and pressure patterns in reservoir around producing well

pressure of 1,900 psi, producing 6-ppg oil, a typical wellhead pressure might be 650 psi (fig. 102). To keep tubing pressure high, formation pressure must be great enough to offset both the pressure drop into the completion zone and the pressure drop from completion zone to wellhead.

In a solution-gas drive, pressure may be quickly depleted. With a gas-cap drive, however, the expansion of the gas cap and its invasion into the production zone may actually increase wellhead pressure. Gas produced with the oil lowers the density of the oil and thus reduces fluid hydrostatic pressure in the tubing (fig. 103). The gas-oil ratio increases, but if the gas is not withdrawn too fast, bottomhole pressure may remain high for a long time.

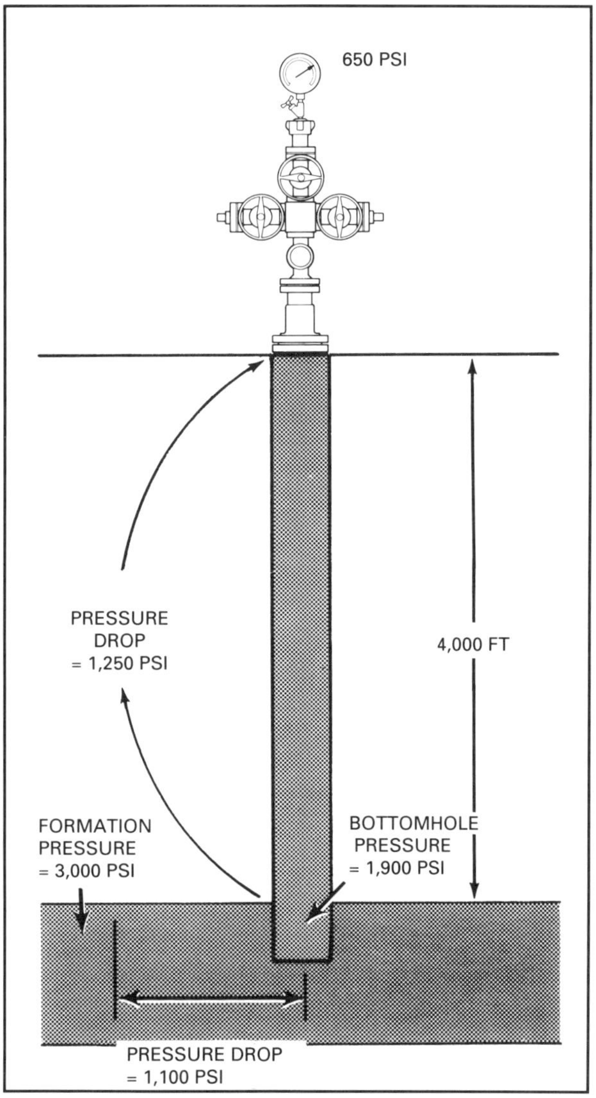

Figure 102. Pressures in a producing well

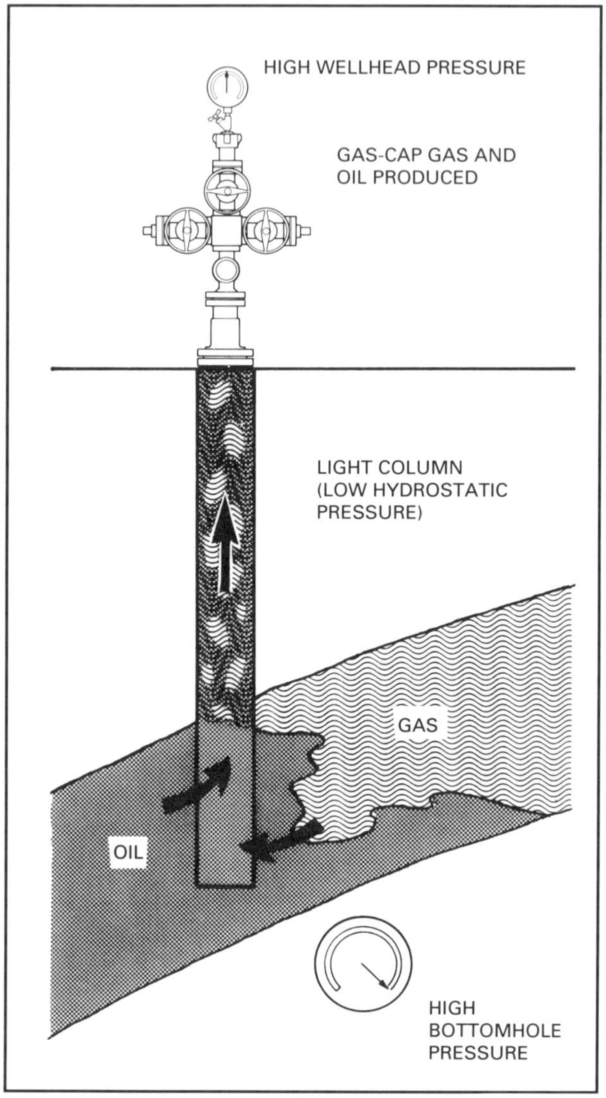

Figure 103. Effect of gas production on tubinghead pressure

Water produced with the oil has the opposite effect—it makes the fluid column heavier, reducing wellhead pressure and slowing oil production (fig. 104). Even though reservoir and bottomhole pressures remain high, artificial lift may be needed.

Gas wells produce from free gas accumulations that may be either associated or nonassociated with oil. Many oilwells produce such high gas-oil ratios that they are classified as gas wells. *Casinghead,* or *associated, gas* is gas produced along with oil, whether from evolved solution gas or from a gas cap. Producing gas directly from the gas cap shrinks the cap and allows it to be invaded by oil, much of which may then adhere to the rock grains and be lost to production. The value

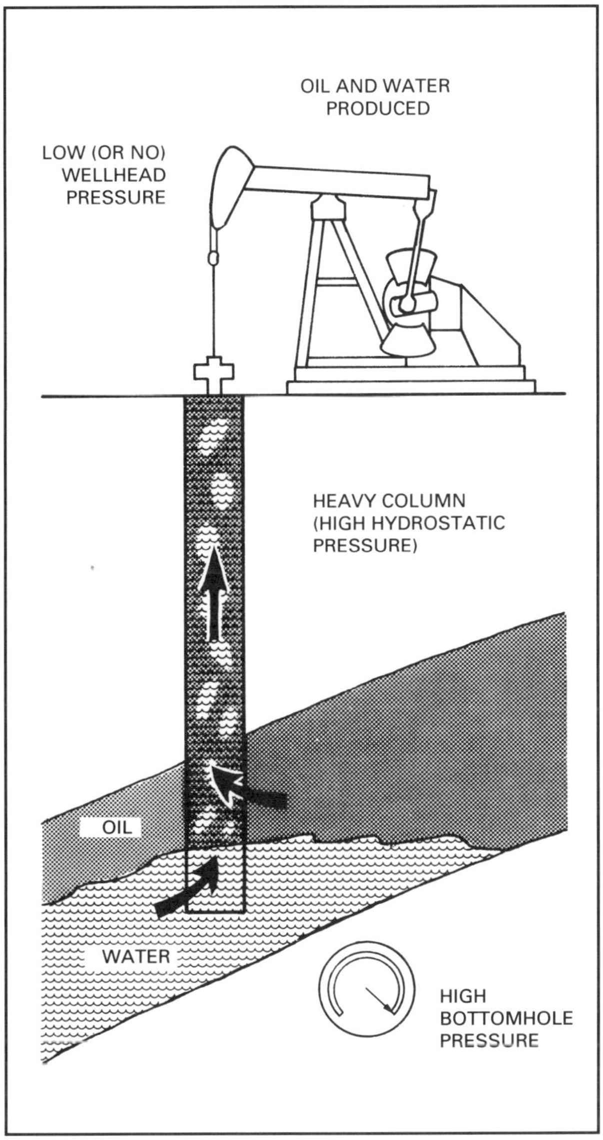

Figure 104. Effect of water production on well pressure

of the oil adsorbed, as well as the oil that cannot be produced because of the loss of drive pressure, usually exceeds the value of the gas produced.

Like oilwells, gas wells must have sufficient pressure to continue to produce hydrocarbons. Bottom- or edgewater may help maintain reservoir pressure. In a depletion drive reservoir, however, gas may arrive at the surface with little pressure remaining and then require recompression for injection into the pipeline system.

An extremely rapid bottomhole pressure drop is sometimes caused by low formation permeability either throughout the reservoir or immediately surrounding the wellbore. Local permeability may be reduced or destroyed during drilling or other operations. Water vapor may condense near the wellbore, where formation pressure and temperature are reduced. This water can reduce permeability by causing clay particles to swell or by simply adhering inside the smaller pore throats and thus increasing capillary pressure. The coning or encroaching of free formation water into a gas well can also reduce permeability. Formation stimulation may be needed to overcome this blockage; otherwise, it may be necessary to shut in or plug the well and either perforate higher in the reservoir or produce from another well higher on the structure.

Reservoir pressure controls not only gas and oil production but also the amount of *condensates* (light hydrocarbon liquids) that are produced with the gas stream. Condensate production drops as pressure drops because, as reservoir pressure drops, some light hydrocarbons condense in the reservoir. But bottomhole permeability or flow problems can cause lower production pressure without reducing the condensate-gas ratio, because reservoir pressure remains high.

REVIEW QUESTIONS
Chapter 7

1. Name three basic requirements a hydrocarbon reservoir must meet to be considered commercially producible.

   A. Must be porous and permeable enough to allow fluids to flow readily into the wellbore.

   B. Proportion of oil and gas must be great enough to ensure that the well is profitable.

   C. Formation pressure must be sufficient to force oil and gas into the wellbore.

D 2. Which drive generally produces the greatest recovery of oil in place?

   A. gas-cap
   B. solution-gas
   C. edgewater
   D. bottom-water

C 3. Gas in a reservoir without oil is called—

   A. solution gas.
   B. casinghead gas.
   C. nonassociated gas.
   D. saturated gas.

B 4. Wellhead pressures are higher in depletion drives than in water drives because—

   A. gas bubbles lighten the well stream.
   B. reservoir pressures are higher.
   C. water drives are less efficient.

5. To be considered commercially producible, a reservoir formation should have at least ___10___% effective porosity.

A 6. In a producing reservoir, pressure is lowest—

   A. around the wellbore.
   B. 10 feet away from the wellbore.
   C. 100 feet away from the wellbore.
   D. in the edgewater.

D 7. In most reservoirs, water is present—

   A. only in the water zone.
   B. only in the water and oil zones.
   C. only in the water and gas zones.
   D. only in the water, oil, and gas zones.

F 8. (T/F) Gravity causes oil and gas to migrate upward.

9. The basic unit of reservoir rock permeability is the Darcy/Millidarcy

B 10. In a solution-gas drive—

A. reservoir pressure rises slowly until most of the oil is gone.
B. reservoir pressure declines rapidly and continuously.
C. reservoir pressure remains relatively constant.

A 11. In a gas-cap drive—

A. the surface gas-oil ratio rises slowly and continuously.
B. reservoir pressure rises rapidly.
C. water production eventually kills the well.

D 12. If produced in excess of a certain rate, a combination drive—

A. becomes a pure bottom-water drive.
B. loads up the well and stops producing.
C. behaves like a pure depletion drive.
D. produces more of the oil in place.

F 13. (T/F) The oil-water contact is a much thinner zone than the gas-oil contact.

14. In a given producing formation, the pressure at some distance from the wellbore is 2,800 psi. If the pressure drop from formation to wellbore is 1,200 psi, and wellhead pressure is 700 psi, what is the pressure drop from the bottom of the hole to the surface? 900 psi

D 15. Low formation permeability near the wellbore can cause—

A. loss of the gas cap.
B. coning of bottom water.
C. excessive wellhead pressure.
D. a very rapid bottomhole pressure drop.

GLOSSARY

A

abnormal pressure *n*: pressure exceeding or falling below the pressure to be expected at a given depth. Normal pressure increases approximately 0.465 pounds per square inch per foot of depth or 10.5 kilopascals per metre of depth. Thus, normal pressure at 1,000 feet is 465 pounds per square inch; at 1,000 metres it is 10,500 kilopascals. See *pressure gradient.*

abrasion *n*: wearing away by friction.

absolute density *n*: the density of a solid or liquid substance at a specified temperature. Sometimes referred to as true density or density in vacuo. See *density.*

absolute open flow *n*: the maximum flow rate that a well could theoretically deliver with zero pressure at the face of the reservoir.

absolute permeability *n*: a measure of the ability of a single fluid (such as water, gas, or oil) to flow through a rock formation when the formation is totally filled (saturated) with that fluid. The permeability measure of a rock filled with a single fluid is different from the permeability measure of the same rock filled with two or more fluids. Compare *permeability, effective permeability, relative permeability.*

absolute porosity *n:* the percentage of the total bulk volume of a rock sample that is composed of pore spaces or voids. Also called total porosity. See *porosity, effective porosity.*

absolute pressure *n*: total pressure measured from an absolute vacuum. It equals the sum of the gauge pressure and the atmospheric pressure. Expressed in pounds per square inch.

absolute viscosity *n*: the property by which a fluid in motion offers resistance to shear and flow. Usually expressed as newton-seconds/metre.

absorption *n*: the process of sucking up, taking in, and making part of an existing whole. Compare *adsorption.*

abyssal *adj*: of or relating to the bottom waters of the ocean.

AC *n*: alternating current.

accumulate *v*: to amass or collect. When oil and gas migrate into porous formations, the quantity collected is called an accumulation.

acoustic log *n*: a record of the measurement of porosity, done by comparing depth to the time it takes for a sonic impulse to travel through a given length of formation. The rate of travel of the sound wave through a rock depends on the composition of the formation and the fluids it contains. Because the type of formation can be ascertained by other logs, and because sonic transit time varies with relative amounts of rock and fluid, porosity can usually be determined in this way.

acoustic well logging *n*: the process of recording the acoustic characteristics of subsurface formations, based on the time required for a sound wave to travel a specific distance through rock. The rate of travel depends on the composition of the formation, its porosity, and its fluid content. Also called sonic logging.

adhesion *n*: a force of attraction that causes molecules of one substance to cling to those of a different substance. Compare *cohesion.*

adsorption *n*: the adhesion of a thin film of a gas or liquid to the surface of a solid. Compare *absorption.*

aeolian *adj*: deposited by wind.

aerobic *adj*: requiring free atmospheric oxygen for normal activity. Compare *anaerobic.*

alluvial fan *n*: a large, sloping sedimentary deposit at the mouth of a canyon, laid down by intermittently flowing water, especially in arid climates, and composed of gravel and sand. The deposit tends to be coarse and unworked, with angular, poorly sorted grains in thin, overlapping sheets. A line of fans may eventually coalesce into an apron that grows broader and higher as the slopes above are eroded away.

anaerobic *adj*: active in the absence of free oxygen. Compare *aerobic.*

angular unconformity *n*: an unconformity in which formations above and below are not parallel. See *unconformity.*

annulus *n*: the space between two concentric circles. In the petroleum industry, it is usually the space surrounding a pipe in the wellbore; sometimes termed the annulus.

anomaly *n*: a deviation from the norm. In geology the term indicates an abnormality such as a fault or dome in a sedimentary bed or a distinctive gravitational or magnetic feature in the crust.

anticlinal trap *n*: a hydrocarbon trap in which petroleum accumulates in the top of an anticline. See *anticline.*

anticline *n*: an arched, or inverted-trough, configuration of folded rock layers. Anticlines sometimes trap oil and gas. See *draped anticline.* Compare *syncline.*

artificial lift *n*: any method used to raise oil to the surface through a well after reservoir pressure has declined to the point at which the well no longer produces by means of natural energy. Sucker rod pumps, gas lift, hydraulic pumps, and submersible electric pumps are the most common means of artificial lift.

associated gas *n*: natural gas that overlies and contacts crude oil in a reservoir. Where reservoir conditions are such that the production of associated gas does not substantially affect the recovery of crude oil in the reservoir, such gas may also be reclassified as nonassociated gas by a regulatory agency. Also called associated free gas. See *gas cap*.

B

barefoot completion *n*: see *open-hole completion*.

barrel (bbl) *n*: 1. a measure of volume for petroleum products in the United States. One barrel is the equivalent of 42 U.S. gallons or 0.15899 cubic metres (9,702 cubic inches). One cubic metre equals 6.2897 barrels.

bar sand *n*: reservoir rock formed from a mass of sand, gravel, or alluvium deposited on the bed of a stream, sea, or lake by waves and currents.

base map *n*: horizontal representation of nongeologic surface features such as streams, roads, buildings, survey benchmarks, and property lines.

basement rock *n*: igneous or metamorphic rock, which seldom contains petroleum. Ordinarily, it lies below sedimentary rock. When it is encountered in drilling, the well is usually abandoned.

basin *n*: 1. a local depression in the earth's crust in which sediments can accumulate to form thick sequences of sedimentary rock. 2. the area drained by a stream and its tributaries. 3. a geologic structure in which strata are inclined toward a common center.

bbl *abbr*: barrel.

bed *n*: a specific layer of earth or rock that presents a contrast to other layers of different material lying above, below, or adjacent to it.

bed load *n*: the gravel and coarse sand that are rolled and bounced along the bottom of a flowing stream. Compare *dissolved load, suspended load*.

benzene *n*: C_6H_6, colorless, volatile, flammable toxic liquid aromatic hydrocarbon used as a solvent and as a motor fuel.

biogenic *adj*: produced by living organisms.

bioherm *n*: a reef or mound built by small organisms and their remains, such as coral, plankton, and oysters. Originally a wave-resistant coral structure served as an anchor for calcareous debris that forms limestone. It was tectonically submerged, or the sea level rose faster than the corals could build it, and it was eventually buried beneath marine shales. A bioherm is often porous enough to hold large accumulations of hydrocarbons, especially if it has been dolomitized. A bioherm is a stratigraphic trap.

bit *n*: the cutting or boring element used in drilling oil and gas wells. The bit consists of a cutting element and a circulating element. The cutting element is steel teeth, tungsten carbide buttons, industrial diamonds, or polycrystalline diamonds (PCDs). These teeth, buttons, or diamonds penetrate and gouge or scrape the formation to remove it. The circulating element permits the passage of drilling fluid and utilizes the hydraulic force of the fluid stream to im-prove drilling rates. In rotary drilling, several drill collars are joined to the bottom end of the drill pipe column, and the bit is attached to the end of the drill collars. Drill collars provide weight on the bit to keep it in firm contact with the bottom of the hole. Most bits used in rotary drilling are roller cone bits, but diamond bits are also used extensively.

bit cone *n*: see *roller cone bit*.

bit drift *n*: the tendency of the bit to move other than vertically, caused by an interaction between the rotation of the bit and the varying resistance of the formation being drilled.

bit walk *n*: the tendency of the bit to drill in the direction of rotation in an inclined hole; e.g., a right-rotating bit walks to the right.

block diagram *n*: three-dimensional geologic representation combining vertical presentation of several sections with the horizontal data normally displayed on a map. The diagram looks like a solid block that has been lifted out of the crust. Compare *panel diagram*.

blowout *n*: an uncontrolled flow of gas, oil, or other well fluids into the atmosphere or into an underground formation. A blowout, or gusher, can occur when formation pressure exceeds the pressure applied to it by the column of drilling fluid. A kick warns of the possibility of a blowout. See *kick*.

blowout preventer *n*: one of several valves installed at the wellhead to prevent the escape of pressure either in the annular space between the casing and drill pipe or in open hole (i.e., hole with no drill pipe) during drilling or completion operations. Blowout preventers on land rigs are located beneath the rig at the land's surface; on jackup or platform rigs, at the water's surface; and on floating offshore rigs, on the seafloor.

bottomhole *n*: the lowest or deepest part of a well. *adj*: pertaining to the bottom of the wellbore.

bottomhole assembly *n*: the portion of the drilling assembly below the drill pipe. It can be very simple—composed of only the bit and drill collars—or it can be very complex and made up of several drilling tools.

bottomhole pressure *n*: 1. the pressure at the bottom of a borehole. It is caused by the hydrostatic pressure of the wellbore fluid and, sometimes, by any back-pressure held at the surface, as when the well is shut in with blowout preventers. When mud is being circulated, bottomhole pressure is the hydrostatic pressure plus the remaining circulating pressure required to move the mud up the annulus. 2. the pressure in a well at a point opposite the producing formation, as recorded by a bottomhole pressure bomb.

bottomset bed *n*: the part of a marine delta that lies farthest from shore. It consists of silt and clay extending well out from the toe of the steep delta face. Such beds grow slowly, out of reach of most of the effects of river current and wave action.

bottom water *n*: interstitial water found directly beneath the oil zone in a petroleum reservoir.

bottom-water drive *n*: see *water drive*.

bottom-water reservoir *n*: a petroleum reservoir in which the pressure of water directly beneath the oil zone drives oil into the well.

breccia *n*: a conglomerate rock composed largely of angular fragments greater than 2 mm in diameter.

buoyancy *n*: the apparent loss of weight of an object immersed in a fluid. If the object is floating, the immersed portion displaces a volume of fluid the weight of which is equal to the weight of the object.

buoyant *adj*: see *buoyancy*.

C

calcareous *adj*: containing or composed largely of calcium carbonate, or calcite ($CaCO_3$).

calcite *n*: calcium carbonate, $CaCO_3$.

calcium *n*: one of the alkaline earth elements with a valence of 2 and an atomic weight of about 40. Calcium compounds are a common cause of water hardness. Calcium is also a component of lime, gypsum, and limestone.

calcium carbonate *n*: a chemical combination of calcium, carbon, and oxygen, $CaCO_3$. It is the main constituent of limestone. It forms a tenacious scale in water-handling facilities and is a cause of water hardness.

caliper log *n*: a record showing variations in wellbore diameter by depth, indicating undue enlargement due to caving in, washout, or other causes. The caliper log also reveals corrosion, scaling, or pitting inside tubular goods.

capillaries *n pl*: very small fissures or cracks in a formation through which water or hydrocarbons flow.

capillarity *n*: the tendency of liquids to rise into small spaces such as rock pores, caused by the combined action of surface tension (cohesion) and wetting (adhesion). See *capillary pressure*.

capillary pressure *n*: a pressure or adhesive force caused by the surface tension of water. This pressure causes the water to adhere more tightly to the surface of small pore spaces than to larger ones. Capillary pressure in a rock formation is comparable to the pressure of water that rises higher in a small glass capillary tube than it does in a larger tube.

caprock *n*: 1. a disklike plate of anhydrite, gypsum, limestone, or sulfur overlying most salt domes in the Gulf Coast region. 2. impermeable rock overlying an oil or gas reservoir that tends to prevent migration of oil or gas out of the reservoir.

carbonate *n*: 1. a salt of carbonic acid. 2. a compound containing the carbonate radical (CO_3). 3. sedimentary rock—limestone or dolomite—that consists mostly of calcium carbonate or magnesium carbonate.

carbonate mud *n*: a mud that forms on the seafloor by the accumulation of calcite particles. It may eventually become limestone.

carbonate reef *n*: see *reef*.

cased *adj*: pertaining to a wellbore in which casing has been run and cemented. See *casing*.

cased hole *n*: a wellbore in which casing has been run.

casing *n*: 1. steel pipe placed in an oil or gas well to prevent the wall of the hole from caving in, to prevent movement of fluids from one formation to another, and to improve the efficiency of extracting petroleum if the well is productive. A joint of casing is available in three length ranges: a joint of range 1 casing is 16 to 25 feet long; a joint of range 2 casing is 25 to 34 feet long; and a joint of range 3 casing is 34 to 48 feet long. Diameters of casing manufactured to API specifications range from 4½ to 20 inches. Casing is also made of many types of steel alloy, which vary in strength, corrosion resistance, and so on.

casinghead gas *n*: unprocessed natural gas produced from a reservoir containing oil.

catastrophism *n*: the theory that the earth's landforms assumed their present configuration in a brief episode at the beginning of geologic history—possibly in a single great catastrophic event—and have remained relatively unchanged since that time. Compare *uniformitarianism*.

cavern *n*: a natural cavity in the earth's crust that is large enough to permit human entry. Commonly formed in limestone by groundwater leaching. Compare *vug*.

cavernous formation *n*: a rock formation that contains large open spaces, usually resulting from the dissolving of soluble substances by formation waters that may still be present. See *vug*.

caving *n*: collapsing of the walls of the wellbore. Also called sloughing.

cavings *n pl*: particles that fall off (are sloughed from) the wall of the wellbore. Compare *cuttings*.

cc *abbr*: 1. cubic centimetre. Also abbreviated cm^3.

cementation *n*: 1. the crystallization or precipitation of soluble minerals in the pore spaces between clastic particles, causing them to become consolidated into sedimentary rock. 2. precipitation of a binding material around grains or minerals in rocks.

Cenozoic era *n*: the time period from 65 million years ago until the present. It is marked by rapid evolution of mammals and birds, flowering plants, grasses, and shrubs, and little change in invertebrates.

centimetre (cm) *n*: a unit of length in the metric system equal to one-hundredth of a metre (10^{-2} metre).

CF *abbr*: cubic foot. Also abbreviated ft^3.

choke *n*: a device with an orifice installed in a line to restrict the flow of fluids. Surface chokes are part of the Christmas tree on a well and contain a choke nipple, or bean, with a small-diameter bore that serves to restrict the flow. Chokes are also used to control the rate of flow of the drilling mud out of the hole when the well is closed in with the blowout preventer and a kick is being circulated out of the hole.

circulate *v*: to pass from one point throughout a system and back to the starting point. For example, drilling fluid is circulated out of the suction pit, down the drill pipe and drill collars, out the bit, up the annulus, and back to the pits while drilling proceeds.

circulation *n*: the movement of drilling fluid out of the mud pits, down the drill stem, up the annulus, and back to the mud pits.

clastic rock *n*: a sedimentary rock composed of fragments of preexisting rocks. The principal distinction among clastics is grain size. Conglomerates, sandstones, and shales are clastic rocks.

clastics *n pl*: 1. sediments formed by the breakdown of large rock masses by climatological processes, physical or chemical. 2. the rocks formed from these sediments.

clastic texture *n*: rock texture in which individual rock, mineral, or organic fragments are cemented together by an amorphous or crystalline mineral such as calcite. Compare *crystalline texture*.

clay *n*: 1. a term used for particles smaller than 1⁄256 millimetre (4 microns) in size, regardless of mineral composition. 2. a group of hydrous aluminum silicate minerals (clay minerals). 3. a sediment of fine clastics.

closure *n*: the vertical distance between the highest point in a petroleum reservoir and its spill point, an indication of the amount of production that may be expected.

cm *abbr*: centimetre.

cm^2 *abbr*: square centimetre.

cm^3 *abbr*: cubic centimetre. Also abbreviated cc.

cohesion *n*: the attractive force between the same kinds of molecules (i.e., the force that holds the molecules of a substance together). Compare *adhesion*.

combination drive *n*: a combination of two or more natural energies that work together in a reservoir to force fluids into a wellbore. Possible combinations include gas-cap and water drive, solution-gas and water drive, and gas-cap drive and gravity drainage.

combination trap *n*: a hydrocarbon trap that has both structural and stratigraphic features.

commercial production *n*: oil and gas production of sufficient quantity to justify keeping a well in production.

compaction *n*: a decrease in the volume of sediments or rock due to pressure exerted by overlying layers, evaporation of water, or other causes.

complete a well *v*: to finish work on a well and bring it to productive status. See *well completion*.

compressibility *n*: the change in volume per unit of volume of a fluid caused by a unit change in pressure at constant temperature.

compression *n*: the act or process of squeezing a given volume of a substance into a smaller space.

condensate *n*: a light hydrocarbon liquid obtained by condensation of hydrocarbon vapors. It consists of varying proportions of butane, propane, pentane, and heavier fractions, with little or no methane or ethane.

condensation *n*: the process by which vapors are converted into liquids, chiefly accomplished by cooling the vapors, lowering the pressure on the vapors, or both.

conductivity *n*: 1. the ability to transmit or convey (as heat or electricity). 2. an electrical logging measurement obtained from an induction survey, in which eddy currents produced by an alternating magnetic field induce in a receiver coil a voltage proportionate to the ability of the formation to conduct electricity.

conductor line *n*: a small-diameter conductive line used in electric wireline operations, such as electric well logging and perforating, in which the transmission of electrical current is required. Compare *wireline*.

cone *n*: a conical-shaped metal device into which cutting teeth are formed or mounted on a roller cone bit. See *roller cone bit*.

cone bit *n*: a roller bit in which the cutters are conical. See *bit*.

conglomerate *n*: sedimentary rock composed of pebbles of various sizes held together by a cementing material such as clay. Conglomerates are similar to sandstone but are composed mostly of grains more than 2 mm in diameter. Most conglomerates are found in discontinuous, thin, isolated layers; they are not very abundant. In common usage, the term "conglomerate" is restricted to coarse sedimentary rock with rounded grains; conglomerates made up of sharp, angular fragments are called *breccia*.

coning *n*: the drawing of reservoir water up, or gas down, into the oil zone of a reservoir by excessively fast production. The water or gas tends to encroach near the wellbore and assume a conical shape.

connate water *n*: water retained in the pore spaces, or interstices, of a formation from the time the formation was created. Compare *interstitial water*.

consolidation *n*: the process by which sand or other loose materials become firm.

contact *n*: 1. in geology, any sharp or well-defined boundary between two different bodies of rock. 2. a bedding plane or unconformity that separates formations.

contact metamorphism *n*: a type of metamorphism that occurs when an intruded body of molten igneous rock changes the rocks immediately around it, primarily by heating and by chemical alteration. See *metamorphism*.

continental crust *n*: that part of the earth's outer layer, varying in thickness from 10 to 30 miles (15-50 km), that makes up the continents. It is composed chiefly of aluminum, silicon, and oxygen and is lighter and thicker than oceanic crust. See *crust, oceanic crust*.

continental drift *n*: according to a theory proposed by Alfred Wegener, a German meteorologist, in 1910, the migration of continents across the ocean floor like rafts drifting at sea. Compare *plate tectonics*.

continental margin *n*: a zone that separates emergent continents from the deep sea bottom, or oceanic abyss.

continental rise *n*: the transition zone between the continental slope and the oceanic abyss.

continental shelf *n*: a zone, adjacent to a continent, that extends from the lower waterline to the continental slope, the point at which the seafloor begins to slope off steeply into the oceanic abyss.

continental slope *n*: a zone of steep, variable topography forming a transition from the continental shelf edge to the ocean basin.

contour map *n*: a map constructed with continuous lines connecting points of equal value, such as elevation, formation thickness, and rock porosity.

convergence zone *n*: a long, narrow zone on the earth's surface where two or more moving crustal plates meet head-on. Usually marked by a long trench, a string of volcanoes, or a mountain chain.

core *n*: 1. a cylindrical sample taken from a formation for geological analysis. 2. the metallic, partly solid, and partly molten interior of the earth, about 4,400 miles in diameter. *v*: to obtain a solid, cylindrical formation sample for analysis.

core analysis *n*: laboratory analysis of a core sample to determine porosity, permeability, lithology, fluid content, angle of dip, geological age, and probable productivity of the formation.

core-drill *v*: to obtain a core from the bottom of an existing well by using a core bit. See *core* (defn. 1).

corer *n*: a tool used to obtain cylindrical samples of rock and other materials from the wellbore.

core sample *n*: 1. a small portion of a formation obtained by using a core bit in an existing wellbore. 2. a spot sample of the contents of an oil or oil product storage tank usually obtained with a thief, or core sampler, at a given height in the tank.

coring *n*: the process of cutting a vertical, cylindrical sample of the formations encountered as an oilwell is drilled. The purpose of coring is to obtain rock samples, or cores, in such a manner that the rock retains the same properties that it had before it was removed from the formation.

correlate *v*: to relate subsurface information obtained from one well to that of others so that the formations may be charted and their depths and thicknesses noted. Correlations are made by comparing electrical well logs, radioactivity logs, and cores from different wells. See *geologic correlation*.

correlation *n*: see *geologic correlation*.

Cretaceous *adj*: of or relating to the geologic period from about 135 million to 65 million years ago at the end of the Mesozoic era, or to the rocks formed during this period, including the extensive chalk deposits for which it was named.

critical angle *n*: in seismology, the angle at which a seismic wave encountering a velocity boundary is neither transmitted through nor reflected away from the boundary, but instead travels along it. See *velocity boundary*.

crooked hole *n*: a wellbore that has been unintentionally drilled in a direction other than vertical. It usually occurs where there is a section of alternating hard and soft strata steeply inclined from the horizontal.

cross-bedding *n*: sedimentation in which laminations are transverse to the main stratification planes.

cross section *n*: a geological or geophysical profile of a vertical section of the earth.

crude oil *n*: unrefined liquid petroleum. It ranges in gravity from 9° API to 55° API and in color from yellow to black, and it may have a paraffin, asphalt, or mixed base. If a crude oil, or crude, contains a sizable amount of sulfur or sulfur compounds, it is called a sour crude; if it has little or no sulfur, it is called a sweet crude. In addition, crude oils may be referred to as heavy or light according to API gravity, the lighter oils having the higher gravities.

crust *n*: the outer layer of the earth, varying in thickness from 5 to 30 miles (7–50 km). It is composed chiefly of aluminum, magnesium, silicon, and oxygen. See *continental crust, oceanic crust*.

crystalline texture *n*: rock texture that is the result of progressive and simultaneous interlocking growth of mineral crystals. Compare *clastic texture*.

crystallization *n*: the formation of crystals from solutions or melts.

cubic centimetre (cm^3) *n*: a commonly used unit of volume measurement in the metric system equal to 10^{-6} cubic metre, or 1 millilitre. The volume of a cube whose edge is one centimetre.

cubic foot (CF or ft^3) *n*: the volume of a cube, all edges of which measure 1 foot. Natural gas in the United States is usually measured in cubic feet, with the most common standard cubic foot being measured at 60° F and 14.65 pounds per square inch absolute, although base conditions vary from state to state.

cuttings *n pl*: the fragments of rock dislodged by the bit and brought to the surface in the drilling mud. Washed and dried cuttings samples are analyzed by geologists to obtain information about the formations drilled.

cuttings-sample log *n*: a record of hydrocarbon content in cuttings gathered at the shale shaker; usually recorded on the mud log.

D

darcy (*pl*, **darcys**) *n*: a unit of measure of permeability. A porous medium has a permeability of 1 darcy when differential pressure of 1 atmosphere across a sample 1 cm long and 1 cm^2 in cross section will force a liquid of 1 centipoise of viscosity through the sample at the rate of 1 cm^3 per second. The permeability of reservoir rocks is usually so low that it is measured in millidarcys (md).

deflection *n*: a change in the angle of a wellbore. In directional drilling, it is measured in degrees from the vertical.

deflection tool *n*: a device made up in the drill string that causes the bit to drill at an angle to the existing hole. It is often called a kickoff tool, because it is used at the kickoff point to start building angle.

deflocculation *n*: the dispersion of solids that have stuck together in drilling fluid, usually by means of chemical thinners.

deformation *n*: the action of earth stresses that results in folding, faulting, shearing, or compression of rocks.

delta *n*: a typically fan-shaped body of sediments, deposited by a river, that projects beyond the adjacent shoreline of a lake or ocean. See *lacustrine delta, marine delta.*

density *n*: the mass or weight of a substance per unit volume. For instance, the density of a drilling mud may be 10 ppg, 74.8 lb/ft^3, or 1,198.2 kg/m^3. Specific gravity, relative density, and API gravity are other units of density.

density log *n*: a special radioactivity log for open-hole surveying that responds to variations in the specific gravity of formations. It is a contact log (i.e., the logging tool is held against the wall of the hole). It emits neutrons and then measures the secondary gamma radiation that is scattered back to the detector in the instrument. The density log is an excellent porosity-measure device, especially for shaly sands. Some trade names for it are Formation Density Log, Gamma-Gamma Density Log, and Densilog.

deplete *v*: to exhaust a supply. An oil and gas reservoir is depleted when most or all economically recoverable hydrocarbons have been produced.

depletion *n*: 1. the exhaustion of a resource. 2. a reduction in income reflecting the exhaustion of a resource. The concept of depletion recognizes that a natural resource such as oil is used up over several accounting periods and permits the value of this resource to be expensed periodically as the resource is exhausted.

depletion drive *n*: a type of reservoir drive in which the expansion of a gas cap or dissolved (solution) gas drives oil into the wellbore and to the surface. See *reservoir drive, gas-cap drive, solution-gas drive.*

deposition *n*: the laying down of sediments or other potential rock-forming material.

depositional energy *n*: the energy characterizing a particular depositional environment. High depositional energy carries away finer sediments, leaving behind coarse materials such as pebbles and gravel that eventually form conglomerates, while lower depositional energies allow finer sediments to settle out, creating the starting materials for sandstones and shales. As a stream flows from highlands toward the sea, it usually passes through successively lower energy environments. A particular depositional environment may also be subject to seasonal fluctuations, accumulating alternating layers of coarser and finer sediments.

depositional environment *n*: the set of physical, chemical, and geologic conditions (such as climate, stream flow, and sediment source) under which a rock layer was laid down.

Devonian *adj*: of or relating to the geologic period from about 400 million to 350 million years ago in the Paleozoic era, or to the rocks formed during this period, including those of Devonshire, England, where outcrops of such rock were first identified.

diagenesis *n*: the chemical and physical changes that sedimentary deposits undergo (compaction, cementation, recrystallization, and sometimes replacement) during and after lithification.

diapir *n*: a dome or anticlinal fold in which a mobile plastic core, usually salt (halite), has ruptured the more brittle overlying rock. Also called piercement dome.

diastrophism *n*: the process or processes of deformation of the earth's crust that produce oceans, continents, mountains, folds, and faults.

differential pressure *n*: the difference between two fluid pressures; for example, the difference between the pressure in a reservoir and in a wellbore drilled in the reservoir, or between atmospheric pressure at sea level and at 10,000 feet. Also called pressure differential.

differential sticking *n*: a condition in which the drill stem becomes stuck against the wall of the wellbore because part of the drill stem (usually the drill collars) has become embedded in the filter cake. Necessary conditions for differential-pressure sticking, or wall sticking, are a permeable formation and a pressure differential across a nearly impermeable filter cake and drill stem. Also called wall sticking. See *filter cake*.

dip *n*: the angle at which a formation bed inclines away from the horizontal. Also used to describe the orientation of a fault.

dip slip *n*: upward or downward displacement along a fault plane.

dip-slip fault *n*: a fault in which the slip is primarily in the line of the fault dip.

disconformity *n*: an interruption in an otherwise continuous succession of sedimentary layers, due to lack of deposition or to erosion, above and below which the strata are parallel to each other (although not necessarily to the disconformity). See *unconformity*; compare *nonconformity*.

displace *v*: 1. to remove physically out of position, as when slippage along a fault displaces rock layers on either side. 2. to drive out, as when water displaces oil in a petroleum reservoir.

displacement *n*: replacement of one fluid by another in the pore space of a reservoir. For example, oil may be displaced by water.

dissolved gas *n*: natural gas that is in solution with crude oil in the reservoir. Compare *free gas*.

dissolved-gas drive *n*: a type of solution-gas drive. See *reservoir drive*.

dissolved load *n*: in a flowing stream of water, those products of weathering that are carried along in solution. Compare *bed load, suspended load*.

dolomitic limestone *n*: a type of sedimentary rock similar to limestone but containing more that 50% magnesium carbonate; sometimes a reservoir rock for petroleum. Often called simply *dolomite*.

dolomitization *n*: the shrinking of the solid volume of rock as limestone turns to dolomite, i.e., the conversion of limestone to dolomite rock by replacement of a portion of the calcium carbonate with magnesium carbonate.

dome *n*: a geologic structure resembling an inverted bowl, i.e., a short anticline that dips or plunges on all sides.

downdip *adj*: lower on the formation dip angle than a particular point.

downhole *adj, adv*: pertaining to the wellbore.

downthrow *n*: the wall of a fault that has moved relatively downward.

draped anticline *n*: an anticline composed of sedimentary deposits atop a reef or atoll, along whose flanks greater thicknesses of sediments have been deposited and compacted than atop the reef itself. Also called compaction anticline.

drift *v*: to move slowly out of alignment, off center, or out of register. See *bit drift*.

drift angle *n*: the angle at which a wellbore deviates from the vertical, expressed in degrees, as revealed by a directional survey. Also called angle of deviation, angle of drift, and inclination.

drill *v*: to bore a hole in the earth, usually to find and remove subsurface formation fluids such as oil and gas.

drill bit *n*: the cutting or boring element used for drilling. See *bit*.

drill collar *n*: a heavy, thick-walled tube, usually steel, placed between the drill pipe and the bit in the drill stem. Several drill collars are used to provide weight on the bit and to provide a pendulum effect to the drill stem. When manufactured to API specifications, a drill collar joint is 30 or 31 feet long.The outside diameter of drill collars made to API specifications ranges from 3⅛ inches to 11 inches.

drill column *n*: see *drill stem*.

driller's log *n*: a record that describes each formation encountered and lists the drilling time relative to depth, usually in 5- to 10-ft (1.5- to 3-m) intervals.

drilling fluid *n*: circulating fluid, one function of which is to lift cuttings out of the wellbore and to the surface. Other functions are to cool the bit and to counteract downhole formation pressure. Although a mixture of barite, clay, water, and chemical additives is the most common drilling fluid, wells can also be drilled by using air, gas, water, or oil-base mud as the drilling mud. See *mud*.

drilling mud *n*: a specially compounded liquid circulated through the wellbore during rotary drilling operations. See *drilling fluid, mud*.

drilling rate *n*: the speed at which the bit drills the formation; usually called the rate of penetration.

drill pipe *n*: seamless steel or aluminum pipe made up in the drill stem between the kelly or top drive on the surface and the drill collars on the bottom. During drilling, it is usually rotated while drilling fluid is circulated through it. Drill pipe joints are available in three ranges of length: 18 to 22 feet, 27 to 30 feet, and 38 to 45 feet. The most popular length is 27 to 30 feet. It is available with outside diameters ranging from 2⅞ to 5½ inches. Several joints are made up (screwed together) to form the drill string.

drill stem *n*: all members in the assembly used for rotary drilling from the swivel to the bit, including the kelly, drill pipe and tool joints, drill collars, stabilizers, and various specialty items. Compare *drill string*.

drill stem test *n*: the conventional method of formation testing. The basic drill stem test tool consists of a packer or packers, valves or ports that may be opened and closed from the surface, and two or more pressure-recording devices. The tool is lowered on the drill string to the zone to be tested. The packer or packers are set to isolate the zone from the drilling fluid column. The valves or ports are then opened to allow for formation flow while the recorders chart static pressures. A sampling chamber traps clean formation fluids at the end of the test. Analysis of the pressure charts is an important part of formation testing.

drill string *n*: the column, or string, of drill pipe with attached tool joints that transmits fluid and rotational power from the kelly to the drill collars and bit. Often, especially in the oil patch, the term is loosely applied to both drill pipe and drill collars. Compare *drill stem.*

drive *n*: see *reservoir drive.*

dry hole *n*: any well that does not produce oil or gas in commercial quantities. A dry hole may flow water, gas, or even oil, but not in amounts large enough to justify production.

E

edgewater *n*: the water that touches the edge of the oil in the lower horizon of a formation.

edgewater drive *n*: see *water drive.*

effective permeability *n*: a measure of the ability of a single fluid to flow through a rock when another fluid is also present in the pore spaces. Compare *absolute permeability, relative permeability, permeability.*

effective porosity *n*: the percentage of the bulk volume of a rock sample that is composed of interconnected pore spaces that allow the passage of fluids through the sample. See *absolute porosity, porosity.*

electric log *n*: see *electric well log.*

electric survey *n*: see *electric well log.*

electric well log *n*: a record of certain electrical characteristics (such as resistivity and conductivity) of formations traversed by the borehole. It is made to identify the formations, determine the nature and amount of fluids they contain, and estimate their depth. Also called an electric log or electric survey.

electrode *n*: a conductor of electric current as it leaves or enters a medium such as an electrolyte, a gas, or a vacuum.

element *n*: one of more than a hundred simple substances that consist of atoms of only one kind and that either singly or in combination make up all matter. For example, the simplest element is hydrogen, and one of the most abundant elements is carbon. Some elements, such as radium and uranium, are radioactive.

encroachment *n*: see *water encroachment.*

entrained *adj*: drawn in and transported by the flow of a fluid.

entrained gas *n*: formation gas that enters the drilling fluid in the annulus.

entrapment *n*: the underground accumulation of oil and gas in geological traps.

environment of deposition *n*: see *depositional environment.*

epeiric sea *n*: a shallow arm of the ocean that extends from the continental shelf deep into the interior of the continent. Also called epicontinental sea.

epicontinental sea *n*: see *epeiric sea.*

epoch *n*: a division of geologic time; a subdivision of a geologic period.

era *n*: one of the major divisions of geologic time.

erosion surface *n*: in subsurface geology, the top surface of a body of rock that has been exposed to and modified by running water or other agents, then buried as an unconformity.

escarpment *n*: a cliff or relatively steep slope that separates level or gently sloping areas of land.

estuary *n*: a coastal indentation or bay into which a river empties and where fresh water mixes with seawater. Compare *marine delta.*

ethane *n*: a paraffin hydrocarbon, C_2H_6; under atmospheric conditions, a gas. One of the components of natural gas.

ethene *n*: see *ethylene.*

ethylene *n*: a chemical compound of the olefin series with the formula C_2H_4. Official name is ethene.

evaporite *n*: a sedimentary rock formed by precipitation of dissolved solids from water evaporating in enclosed basins. Examples are gypsum and salt.

F

F *abbr.*: degree Fahrenheit. See *Fahrenheit scale.*

facies *n*: part of a bed of sedimentary rock that differs significantly from other parts of the bed.

Fahrenheit scale *n*: a temperature scale devised by Gabriel Fahrenheit, in which 32° represents the freezing point and 212° the boiling point of water at standard sea-level pressure. Fahrenheit degrees may be converted to Celsius degrees by using the following formula: °C = 5/9 (°F-32).

fault *n*: a break in the earth's crust along which rocks on one side have been displaced (upward, downward, or laterally) relative to those on the other side.

fault dip *n*: the vertical inclination of a fault's surface, or plane, measured from a horizontal plane.

fault plane *n*: a surface along which faulting has occurred.

fault trap *n*: a subsurface hydrocarbon trap created by faulting, in which an impermeable rock layer has moved opposite the reservoir bed or where impermeable gouge has sealed the fault and stopped fluid migration.

faunal succession *n*: the principle that fossils in a stratigraphic sequence succeed one another in a definite, recognizable order.

feldspar *n*: a group of silicate minerals that includes a wide variety of potassium, sodium, and aluminum silicates. Feldspar makes up about 60% of the outer 15 km of the earth's crust.

fence diagram *n*: see *panel diagram*.

field *n*: a geographical area in which a number of oil or gas wells produce from a continuous reservoir. A field may refer to surface area only or to underground productive formations as well. A single field may have several separate reservoirs at varying depths.

filter cake *n*: the layer of concentrated solids from the drilling mud or cement slurry that forms on the walls of the borehole opposite permeable formations; also called wall cake or mud cake.

filtration loss *n*: the escape of the liquid part of a drilling mud into permeable formations.

fishing *n*: the procedure of recovering lost or stuck equipment in the wellbore.

fissure *n*: a crack or fracture in a subsurface formation.

flowing well *n*: a well that produces oil or gas by its own reservoir pressure rather than by use of artificial means (such as pumps).

fluid *n*: a substance that flows and yields to any force tending to change its shape. Liquids and gases are fluids, as are many solids, such as rock, under extreme pressure and/or temperature.

fluid contact *n*: the approximate point in a reservoir where the gas-oil contact or oil-water contact is located.

fluvial *adj*: deposited by flowing water, such as a stream.

fold *n*: a flexure of rock strata (e.g, an arch or a trough) produced by horizontal compression of the earth's crust. See *anticline, syncline*.

fold belt *n*: any region of the earth's crust characterized by extensive flexure of rock strata into anticlines and synclines.

footwall *n*: the rock surface forming the underside of a fault when the fault plane is not vertical—that is, if the dip is less than 90°. Compare *hanging wall*.

foraminifera *n pl*: single-celled, mostly microscopic animals with calcareous exoskeletons; mostly marine.

foreset bed *n*: a depositional layer on the steep seaward face of a marine delta that lies beyond the topset beds and is composed of finer sedimentary materials than the topset beds.

foreshore *n*: that part of the seashore that lies between low- and high-tide levels.

formation *n*: a bed or deposit composed throughout of substantially the same kind of rock; often a lithologic unit. Each formation is given a name, frequently as a result of the study of the formation outcrop at the surface and sometimes based on fossils found in the formation.

formation boundary *n*: the horizontal limits of a formation.

formation dip *n*: the angle at which a formation bed inclines away from the horizontal. Dip is also used to describe the orientation of a fault.

formation evaluation *n*: the analysis of subsurface formation characteristics, such as lithology, porosity, permeability, and saturation, by indirect methods such as wireline well logging or by direct methods such as mud logging and core analysis.

formation face *n*: that part of a formation exposed to the wellbore.

formation fluid *n*: fluid (such as gas, oil, or water) that exists in a subsurface rock formation.

formation fracture gradient *n*: a plot of pressure versus depth that reveals the pressure at which a formation will fracture at a given depth.

formation fracture pressure *n*: the point at which a formation will crack from pressure in the wellbore.

formation fracturing *n*: a method of stimulating production by opening new flow channels in the rock surrounding a production well. Often called a frac job. Under extremely high hydraulic pressure, a fluid (such as distillate, diesel fuel, crude oil, dilute hydrochloric acid, water, or kerosene) is pumped downward through production tubing or drill pipe and forced out below a packer or between two packers. The pressure causes cracks to open in the formation, and the fluid penetrates the formation through the cracks. Sand grains, aluminum pellets, walnut shells, or similar materials (propping agents) are carried in suspension by the fluid into the cracks. When the pressure is released at the surface, the fracturing fluid returns to the well. The cracks partially close on the pellets, leaving channels for oil to flow around them to the well.

formation gas *n*: gas initially produced from an underground reservoir.

formation pressure *n*: the force exerted by fluids in a formation, recorded in the hole at the level of the formation with the well shut in. Also called reservoir pressure or shut-in bottomhole pressure.

formation resistivity *n*: a measure of the electrical resistance of fluids in a formation.

formation strength *n*: the ability of a formation to resist fracture from pressures created by fluids in a borehole.

formation testing *n*: the gathering of pressure data and fluid samples from a formation to determine its production potential before choosing a completion method. Formation testing tools include formation testers and drill stem test tools.

formation water *n*: 1. the water originally in place in a formation. See *connate water*. 2. any water that resides in the pore spaces of a formation.

fossil *n*: the remains or impressions of a plant or animal of past geological ages that have been preserved in or as rock.

fossiliferous *adj*: containing fossils.

fracture *n*: a crack or crevice in a formation, either natural or induced.

fracture pressure *n*: the pressure at which a formation will break down, or fracture.

fracture zone *n*: zone of naturally occurring fissures or fractures that can pose problems with lost circulation.

free gas *n*: a hydrocarbon that exists in the gaseous phase at reservoir pressure and temperature and remains a gas when produced under normal conditions. Compare *dissolved gas*.

free water *n*: water produced with oil. It usually settles out within five minutes when the well fluids become stationary in a settling space within a vessel.

fresh water *n*: water that has little or no salt dissolved in it.

frost heaving *n*: movement of the soil resulting from alternate thawing and freezing. Frost heaving generates stress on vertical support members of pipelines in the Arctic and, by extension, on the pipe itself.

frost wedging *n*: the phenomenon resulting when water invades rock, freezes, and, by its expansion, wedges apart the rock. Repeated freeze-thaw cycles can quickly break up any rock that has even the tiniest cracks.

ft *abbr*: foot.

ft² *abbr*: square foot.

ft³ *abbr*: cubic foot.

G

g *abbr*: gram.

gamma particle *n*: a short, highly penetrating X-ray emitted by radioactive substances during their spontaneous disintegration. The measurement of gamma particles (sometimes called gamma rays) is the basis for a number of radioactivity well logging methods.

gamma ray *n*: see *gamma particle*.

gamma ray log *n*: a type of radioactivity well log that records natural radioactivity around the wellbore. Shales generally produce higher levels of gamma radiation and can be detected and studied with the gamma ray tool. In holes where salty drilling fluids are used, electric logging tools are less effective than gamma ray tools. See *radioactivity log*.

gas *n*: a compressible fluid that completely fills any container in which it is confined. Technically, a gas will not condense when it is compressed and cooled, because a gas can exist only above the critical temperature for its particular composition. Below the critical temperature, this form of matter is known as a vapor, because liquid can exist and condensation can occur. Sometimes the terms "gas" and "vapor" are used interchangeably. The latter, however, should be used for those streams in which condensation can occur and that originate from, or are in equilibrium with, a liquid phase.

gas cap *n*: a free-gas phase overlying an oil zone and occurring within the same producing formation as the oil. See *associated gas, reservoir*.

gas-cap drive *n*: drive energy supplied naturally (as a reservoir is produced) by the expansion of gas in a cap overlying the oil in the reservoir. In an oil reservoir, the energy, or drive, provided by free gas that lies above the oil. In such a drive, the gas cap expands to force oil into the well and to the surface. See *reservoir drive*.

gas-cap gas *n*: see *associated gas*.

gas drive *n*: the use of the energy that arises from the expansion of compressed gas in a reservoir to move crude oil to a wellbore. Also called depletion drive. See *reservoir drive*.

gas expansion *n*: when oil and gas are found in the same reservoir under pressure, the drilling of a well into the reservoir releases the pressure, causing the gas to expand. The expanding gas drives the oil toward and up the wellbore. The expansive energy of the gas can be harnessed whether the gas is in solution or forming a cap above the oil.

gas field *n*: a district or area from which natural gas is produced.

gas-oil contact *n*: the point or plane in a reservoir at which the bottom of a gas sand is in contact with the top of an oil sand.

gas-oil ratio *n*: a measure of the volume of gas produced with oil, expressed in cubic feet per barrel or cubic metres per tonne. Also called solution gas-oil ratio.

gas reservoir n: a geological formation containing a single gaseous phase. When produced, the surface equipment may or may not contain condensed liquid, depending on the temperature, pressure, and composition of the single reservoir phase.

gas sand *n*: a stratum of sand or porous sandstone from which natural gas is obtained.

gasser *n*: a well that produces natural gas.

gas well *n*: a well that primarily produces gas. Legal definitions vary among the states.

gas zone *n*: an area in a reservoir that is occupied by natural gas.

gel *n*: a semisolid, jellylike state assumed by some colloidal dispersions at rest. When agitated, the gel converts to a fluid state. Also a nickname for bentonite. *v*: to take the form of a gel; to set.

gel strength *n*: a measure of the ability of a colloidal dispersion to develop and retain a gel form, based on its resistance to shear. The gel, or shear, strength of a drilling mud determines its ability to hold solids in suspension. Sometimes bentonite and other colloidal clays are added to drilling fluid to increase its gel strength.

geochemistry *n*: study of the relative and absolute abundances of the elements of the earth and the physical and chemical processes that have produced their observed distributions.

geologic correlation *n*: the relating of stratigraphic information obtained from one well or surface location to that of others.

geologic time scale *n*: the long periods of time dealt with and identified by geology. Geologic time is divided into eras (usually Cenozoic, Mesozoic, Paleozoic, and Precambrian), which are subdivided into periods and epochs. When the age of a type of rock is determined, it is assigned a place in the scale and thereafter referred to as, for example, Mesozoic rock of the Triassic period.

geologist *n*: a scientist who gathers and interprets data pertaining to the rocks of the earth's crust.

geology *n*: the science of the physical history of the earth and its life, especially as recorded in the rocks of the crust.

geomorphic unit *n*: one of the features that, taken together, make up the form of the surface of the earth.

geophone *n*: an instrument placed on the surface that detects vibrations passing through the earth's crust. It is used in conjunction with seismography. Geophones are often called jugs. See *seismograph*.

geophysical exploration *n*: measurement of the physical properties of the earth to locate subsurface formations that may contain commercial accumulations of oil, gas, or other minerals; to obtain information for the design of surface structures; or to make other practical applications. The properties most often studied in the oil industry are seismic characteristics, magnetism, and gravity.

geophysicist *n*: one who studies geophysics.

geophysics *n*: the physics of the earth, including meteorology, hydrology, oceanography, seismology, vulcanology, magnetism, and radioactivity.

geopressure *n*: abnormally high pressure exerted by some subsurface formations. The deeper the formation, the higher the pressure it exerts on a wellbore drilled into it. See *geostatic pressure*.

geopressured shales *n*: impermeable shales, highly compressed by overburden pressure, that are characterized by large amounts of formation fluids and abnormally high pore pressure.

geoscience *n*: a science dealing with the earth—geology, physical geography, geophysics, geomorphology, geochemistry.

geostatic pressure *n*: the pressure to which a formation is subjected by its overburden. Also called ground pressure, lithostatic pressure, rock pressure, overburden pressure. A reservoir that is totally isolated from the surface by impermeable rock is often considered under geostatic pressure (geopressured). Compare *hydrostatic pressure*.

geostatic pressure gradient *n*: the change in geostatic pressure per unit of depth in the earth.

geothermal *adj*: pertaining to heat within the earth.

geothermal gradient *n*: the increase in the temperature of the earth with increasing depth. It averages about 1°F/60 ft, but may be considerably higher or lower.

geothermal reservoir *n*: 1. a subsurface layer of rock containing steam or hot water that is trapped in the layer by overlying impermeable rock. 2. a subsurface layer of rock that is hot but contains little or no water. Geothermal reservoirs are a potential source of energy.

glacial *adj*: relating to or derived from glaciers. Glacial sediments are those created or deposited by glaciers. Often refers to sediments or sedimentary rocks derived from, or surface features created by, widespread glacial activity during an ice age.

Gondwanaland *n*: the southern part of the supercontinent Pangaea, comprising the future land masses of South America, Africa, Antarctica, Australia, and India.

gone to water *adj*: pertaining to a well in which production of oil has decreased and production of water has increased (e.g., "The well has gone to water").

GOR *abbr*: gas-oil ratio.

gouge *n*: finely abraded material occurring between the walls of a fault as the result of grinding movement.

graben *n*: a block of the earth's crust that has subsided between two faults. Compare *horst*.

graded *adj*: characterized by grading. See *grading*.

grading *n*: in a sedimentary unit, a continuous change in texture from coarse sediments on the bottom to fine sediments at the top, caused by a gradual decrease in the depositional energy level.

gradualism *n*: see *uniformitarianism*.

gram *n*: a unit of metric measure of mass and weight equal to 1/1,000 kilogram and nearly equal to 1 cubic centimetre of water at its maximum density.

granite *n*: an igneous rock composed primarily of feldspar, quartz, and mica. It is the most common intrusive rock—that is, it originally solidified below the surface of the earth. Its rock crystals are easily seen by the unaided eye.

gravel pack *n*: a mass of very fine gravel placed around a slotted liner in a well. *v*: to place a slotted or perforated liner in a well and surround it with gravel. See *gravel packing*.

gravel packing *n*: a method of well completion in which a slotted or perforated liner, often wire-wrapped, is placed in the well and surrounded by gravel. If open hole, the well is sometimes enlarged by underreaming at the point where the gravel is packed. The mass of gravel excludes sand from the wellbore but allows continued production.

gravimeter *n*: an instrument used to detect and measure minute differences in the earth's gravitational pull at different locations to obtain data about subsurface formations.

gravimetric survey *n*: the survey made with a gravimeter.

gravity drainage *n*: the movement of fluids in a reservoir resulting from the force of gravity. In the absence of an effective water or gas drive, gravity drainage is an important source of energy to produce oil, and it may also supplement other types of natural drive. Also called segregation drive.

growth fault *n*: an active fault that continues to slip while sediments are being deposited, causing the strata on the downthrust side to be thicker than those on the other side. Also called rollover fault.

gypsum *n*: a naturally occurring crystalline form of calcium sulfate in which each molecule of calcium sulfate is combined with two molecules of water.

H

halite *n*: rock salt (NaCl).

hanging wall *n*: the rock surface forming the upper side of a fault when the fault plane is not vertical—that is, if the dip is less than 90°. Compare *footwall*.

horizon *n*: distinct layer or group of layers of rock.

horizontal permeability *n*: the permeability of reservoir rock parallel to the bedding plane. Compare *vertical permeability*.

horsepower *n*: a unit of measure of work done by a machine. One horsepower equals 33,000 foot-pounds per minute.

horst *n*: a block of the earth's crust that has been uplifted (relatively) between two faults. Compare *graben*.

hydraulic *adj*: 1. of or relating to water or other liquid in motion. 2. operated, moved, or affected by water or liquid.

hydraulic horsepower *n*: a measure of the power of a fluid under pressure.

hydraulics *n*: 1. the branch of science that deals with practical applications of water or other liquid in motion. 2. the planning and operation of a rig hydraulics program, coordinating the power of the circulating fluid at the bit with other aspects of the drilling program so that bottomhole cleaning is maximized.

hydrocarbons *n pl*: organic compounds of hydrogen and carbon whose densities, boiling points, and freezing points increase as their molecular weights increase. Although composed of only two elements, hydrocarbons exist in a variety of compounds, because of the strong affinity of the carbon atom for other atoms and for itself. The smallest molecules of hydrocarbons are gaseous; the largest are solids. Petroleum is a mixture of many different hydrocarbons.

hydrodynamic trap *n*: a petroleum trap in which the major trapping mechanism is the force of moving water.

hydrophone *n*: a device trailed in an array behind a boat in offshore seismic exploration that is used to detect sound reflections, convert them to electric current, and send them through a cable to recording equipment on the boat.

hydrostatic pressure *n*: the force exerted by a body of fluid at rest. It increases directly with the density and the depth of the fluid and is expressed in pounds per square inch or kilopascals. The hydrostatic pressure of fresh water is 0.433 pounds per square inch per foot of depth (9.792 kilopascals per metre). In drilling, the term refers to the pressure exerted by the drilling fluid in the wellbore. In a water drive field, the term refers to the pressure that may furnish the primary energy for production. Hydrostatic pressure is characteristic of reservoirs that communicate with the surface and is considered normal reservoir pressure. Compare *geostatic pressure*.

I

igneous rock *n*: a rock mass formed by the solidification of magma within the earth's crust or on its surface. It is classified by chemical composition and grain size. Granite is an igneous rock.

impermeable *adj*: preventing the passage of fluid. A formation may be porous yet impermeable if there is an absence of connecting passages between the voids within it. See *permeability*.

in. *abbr*: inch.

in.² *abbr*: square inch.

in.³ *abbr*: cubic inch.

inclination n: see *drift angle*.

induction log *n*: see *induction survey*.

induction survey *n*: an electric well log in which the conductivity of the formation rather than the resistivity is measured. Because oil-bearing formations are less conductive of electricity than water-bearing formations, an induction survey, when compared with resistivity readings, can aid in determination of oil and water zones.

initial potential *n*: the early production of an oilwell, recorded after testing operations and recovery of load oil and used as an indicator of the maximum ability of a well to produce on completion without subsequent reservoir damage.

insert bit *n*: see *tungsten carbide bit*.

interdeltaic *adj*: between deltas, as an interdeltaic shoreline.

interface *n*: the contact surface between two boundaries of liquids (e.g., the surface between water and oil in a water-drive reservoir).

interstice *n*: a pore space in a reservoir rock.

interstitial water *n*: water contained in the interstices, or pores, of reservoir rock. In reservoir engineering, it is synonymous with connate water. Compare *connate water*.

intrusive rock *n*: igneous rock that, while molten, penetrated into or between other rocks and solidified.

invaded zone *n*: an area within a permeable rock adjacent to a wellbore into which a filtrate (usually water) from the drilling mud has passed, with consequent partial or total displacement of the fluids originally present in the zone.

isopach map *n*: a geological map of subsurface strata showing the various thicknesses of a given formation as a series of contours. It is widely used in calculating reserves and in planning improved recovery projects.

isopachous line *n*: a contour line drawn on a map joining points of equal thickness in a stratigraphic unit.

isostasy *n*: equilibrium between large segments of the earth's crust, which float on the denser mantle in such a way that thicker segments extend higher and deeper than thinner segments, and lighter blocks rise higher than denser blocks.

J

jet bit *n*: a drilling bit having replaceable nozzles through which the drilling fluid is directed in a high-velocity stream to the bottom of the hole to improve the efficiency of the bit. See *bit*.

joule (J) *n*: the unit used to measure heat, work, and energy in the metric system. It is the amount of energy required to move an object of 1 kilogram mass to a height of 1 metre. Also called a newton-metre.

K

kelvin (K) *n*: the fundamental unit of thermodynamic temperature in the metric system. See *kelvin temperature scale*.

kelvin temperature scale *n*: a temperature scale with the degree interval of the Celsius scale and the zero point at absolute zero. On the kelvin scale, water freezes at 273.16° and boils at 373.16°.

kick *n*: an entry of water, gas, oil, or other formation fluid into the wellbore during drilling. It occurs because the pressure exerted by the column of drilling fluid is not great enough to overcome the pressure exerted by the fluids in the formation drilled. If prompt action is not taken to control the kick, or kill the well, a blowout may occur.

kill *v*: 1. in drilling, to control a kick by taking suitable preventive measures (e.g., to shut in the well with the blowout preventers, circulate the kick out, and increase the weight of the drilling mud). 2. in production, to stop a well from producing oil and gas so that reconditioning of the well can proceed. Production is stopped by circulating a kill fluid into the hole. 3. in production, to stop a flowing well from producing by encroachment of water into the wellbore, making the wellstream too heavy to flow on its own.

kilogram *n*: the metric unit of mass equal to 1,000 grams.

kilometre (km) *n*: a metric unit of length equal to 1,000 metres. One kilometre equals .62 miles.

km *abbr*: kilometre.

L

L *abbr*: litre.

lacustrine *adj*: of or relating to a lake. Lacustrine sediments are those that were deposited in a lake (usually referring to a body of fresh water).

lacustrine delta *n*: a collection of sediment in a lake at the point at which a river or stream enters. When the flowing water enters the lake, the encounter with still water absorbs most or all of the stream's energy, causing its sediment load to be deposited.

landform *n*: a recognizable, naturally formed physical land feature having a characteristic shape, such a plain, alluvial fan, valley, hill, or mountain.

Landsat *n*: an unstaffed earth-orbiting NASA satellite that transmits multispectral images to earth receiving stations; formerly called ERTS (Earth Resource Technology Satellite).

lateral fault *n*: a fault the opposite sides of which move horizontally past each other along a plane that may be vertical; for example, the San Andreas fault. Sometimes called a strike-slip fault.

lateral focus log *n*: a resistivity log taken with a sonde that focuses an electrical current laterally, away from the wellbore, and into the formation being logged. Allows more precise measurement than was possible with earlier sondes. The lateral focus log is a useful means of distinguishing thin rock layers.

Laurasia *n*: the northern part of the supercontinent Pangaea, comprising the future land masses of North America and Asia.

lava *n*: magma that reaches the surface of the earth.

layer *n*: a bed, or stratum, of rock.

lb *abbr*: pound.

leaching *n*: in geology, the removal of minerals from rock by solution in water or another solvent.

lens *n*: 1. a porous, permeable, irregularly shaped sedimentary deposit surrounded by impervious rock. 2. a lenticular sedimentary bed that pinches out, or comes to an end, in all directions.

lens trap *n*: a hydrocarbon reservoir consisting of a porous, permeable, irregularly shaped sedimentary deposit surrounded by impervious rock. See *lens*.

light hydrocarbons *n pl*: the low molecular weight hydrocarbons such as methane, ethane, propane, and butane.

lime mud *n*: 1. a calcite-rich sediment that may give rise to shaly limestone. 2. a drilling mud that is treated with lime to provide a source of soluble calcium in the filtrate to obtain desirable mud properties for drilling in shale or clay formations.

limestone *n*: a sedimentary rock rich in calcium carbonate that sometimes serves as a reservoir rock for petroleum.

lineament *n*: a linear topographic or tonal feature on the terrain and on images and maps of the terrain that is thought to indicate a zone of subsurface structural weakness.

liquid *n*: a state of matter in which the shape of the given mass depends on the containing vessel, but the volume of the mass is independent of the vessel. A liquid is a fluid that is almost incompressible.

lithification *n*: the conversion of unconsolidated deposits into solid rock, usually by burial, compaction, cementation, and diagenesis.

lithofacies map *n*: a facies map showing lithologic variations within a formation. It shows the variations of selected lithologic characteristics within a stratigraphic unit.

lithology *n*: 1. the study of rocks, usually macroscopic. 2. the individual character of a rock in terms of mineral composition, structure, and so forth.

lithostatic pressure *n*: see *geostatic pressure*.

litre (L) *n*: a unit of metric measure of capacity equal to the volume occupied by 1 kilogram of water at 4° C and at the standard atmospheric pressure of 760 millimetres.

loess *n*: unstratified, homogeneous accumulation of silt, often containing small amounts of clay or sand and redeposited by wind from glacial outwash or deserts.

log *n*: a systematic recording of data, such as a driller's log, mud log, electrical well log, or radioactivity log. Many different logs are run in wells to discern various characteristics of downhole formation. *v*: to record data.

log a well *v*: to run any of the various logs used to ascertain downhole information about a well.

logging devices *n pl*: any of several electrical, acoustical, mechanical, or radioactivity devices that are used to measure and record certain characteristics or events that occur in a well that has been or is being drilled.

lost circulation *n*: the quantities of whole mud lost to a formation, usually in cavernous, fissured, or coarsely permeable beds. Evidenced by the complete or partial failure of the mud to return to the surface as it is being circulated in the hole. Lost circulation can lead to a blowout and, in general, can reduce the efficiency of the drilling operation. Also called lost returns.

lost returns *n*: see *lost circulation*.

M

m *sym*: metre.

m² *abbr*: square metre.

m³ *abbr*: cubic metre.

magma *n*: the hot fluid matter within the earth's crust that is capable of intrusion or extrusion and that produces igneous rock when cooled.

magnetic contour map *n*: a geophysical map based on magnetic survey data, on which locations of equal magnetic field strength are connected by contour lines to indicate magnetic anomalies. See *magnetic survey*.

magnetic survey *n*: an exploration method in which an instrument that measures the local strength of the earth's magnetic field is passed over the surface or through the water. The measurements are used to estimate the subsurface depth of dense magnetic basement rock, and thus the thickness of overlying sedimentary layers that may contain oil and gas.

mantle *n*: the hot, plastic part of the earth that lies between the core and the crust. It begins 5 to 30 miles (10 to 50 kilometres) beneath the surface and extends to 1,800 miles (2,900 kilometres).

map *n*: a horizontal representation, usually on a plane surface, of the physical or other features of a surface of the earth, using a specified scale, method of projection, and set of symbols to show the relative size and position of these features.

marine delta *n*: 1. a triangular sea-level extension of land shaped like the Greek letter Δ. 2. a depositional environment in which river-borne sediments accumulate as the flow energy of the river is dissipated in the ocean.

marker bed *n*: a distinctive, easily identified rock stratum, especially one used as a guide for drilling or correlation of logs.

md *abbr*: millidarcy or millidarcys.

Mesozoic era *n*: a span from 230 to 65 million years ago, the era of the dinosaurs and the first mammals.

metamorphic rock *n*: a rock derived from preexisting rocks by mineralogical, chemical, and structural alterations caused by heat and pressure within the earth's crust. Marble is a metamorphic rock.

metamorphism *n*: the process in which rock may be changed by heat and pressure into different forms.

methane *n*: a light, gaseous, flammable paraffinic hydrocarbon, CH_4, that has a boiling point of -25°F and is the chief component of natural gas and an important basic hydrocarbon for petrochemical manufacture.

metre (m) *n*: the fundamental unit of length in the metric system. It is equal to about 3.28 feet, 39.37 inches, or 100 centimetres.

midocean ridge *n*: a sinuous subsea ridge where magma rises to the surface of the oceanic crust and solidifies to form new crust between tectonic plates that are moving apart.

migration *n*: the movement of oil from the area in which it was formed to a reservoir rock where it can accumulate.

millidarcy (md) (*pl*, **millidarcys**) *n*: a measure of permeability equivalent to 0.001 darcy.

millilitre *n*: one-thousandth of a litre. In analyzing drilling mud, this term is used interchangeably with cubic centimetre. A quart equals 964 millilitres.

millimetre *n*: a measurement unit in the metric system equal to 10^{-3} metre (0.001 metre). It is used to measure pipe and bit diameter, nozzle size, liner length and diameter, and cake thickness.

min *abbr*: minute.

mineral *n*: 1. a naturally occurring inorganic crystalline element or compound with a definite chemical composition and characteristic physical properties such as crystal shape, melting point, color, and hardness. Most minerals found in rocks are not pure. 2. broadly, a naturally occurring homogeneous substance that is obtained from the ground for human use (e.g., stone, coal, salt, sulfur, sand, petroleum, water, natural gas).

ml *abbr*: millilitre.

mm *abbr*: millimetre.

molecular weight *n*: the sum of the atomic weights in a molecule. For example, the molecular weight of water, H_2O, is 18, because the atomic weight of each of the hydrogen molecules is 1 and the atomic weight of oxygen is 16.

molecule *n*: the smallest particle of a substance that retains the properties of the substance. It is composed of one or more atoms.

monocline *n*: a steplike bend in otherwise horizontal or gently dipping strata. Compare *anticline, syncline.*

mud *n*: the liquid circulated through the wellbore during rotary drilling and workover operations. In addition to its function of bringing cuttings to the surface, drilling mud cools and lubricates the bit and drill stem, protects against blowouts by holding back subsurface pressures, and deposits a mud cake on the wall of the borehole to prevent loss of fluids to the formation. Although it was originally a suspension of earth solids (especially clays) in water, the mud used in modern drilling operations is a more complex, three-phase mixture of liquids, reactive solids, and inert solids. The liquid phase may be fresh water, diesel oil, or crude oil and may contain one or more conditioners. See *drilling fluid.*

mud cake *n*: the sheath of mud solids that forms on the wall of the hole when liquid from mud filters into the formation. Also called filter cake or wall cake.

mud logger *n*: an employee of a mud logging company who performs mud logging.

mud logging *n*: the recording of information derived from examination and analysis of formation cuttings made by the bit and of mud circulated out of the hole. A portion of the mud is diverted through a gas-detecting device. Cuttings brought up by the mud are examined under ultraviolet light to detect the presence of oil or gas. Mud logging is often carried out in a portable laboratory set up at the well.

mudstone *n*: 1. a massive, blocky rock composed of approximately equal proportions of clay and silt, but lacking the fine lamination of shale. 2. in general, rock consisting of an indefinite and variable mixture of clay, silt, and sand particles.

mud weight *n*: a measure of the density of a drilling fluid expressed as pounds per gallon, pounds per cubic foot, or kilograms per cubic metre. Mud weight is directly related to the amount of pressure the column of drilling mud exerts at the bottom of the hole.

N

natural gas *n*: a highly compressible, highly expansible mixture of hydrocarbons with a low specific gravity and occurring naturally in a gaseous form. Besides hydrocarbon gases, natural gas may contain appreciable quantities of nitrogen, helium, carbon dioxide, hydrogen sulfide, and water vapor. Although gaseous at normal temperatures and pressures, the gases making up the mixture that is natural gas are variable in form and may be found either as gases or as liquids under suitable conditions of temperature and pressure.

natural gas liquids (NGL) *n pl*: those hydrocarbons liquefied at the surface in field facilities or in gas processing plants. Natural gas liquids include propane, butane, and natural gasoline.

neutron log *n*: a radioactivity well log used to determine formation porosity. The logging tool bombards the formation with neutrons. When the neutrons strike hydrogen atoms in water or oil, gamma rays are released. Since water or oil exists only in pore spaces, a measurement of the gamma rays indicates formation porosity. See *radioactivity log*.

nonassociated gas *n*: gas in a reservoir that contains no oil.

nonconformity *n*: a buried landscape in which sediments were deposited on an eroded surface of igneous or metamorphic rock. Compare *unconformity*.

nonporous *adj*: containing no interstices; having no pores.

normal fault *n*: a dip-slip fault along which the hanging wall has subsided relative to the footwall.

normal formation pressure *n*: formation fluid pressure equivalent to about 0.465 pounds per square inch per foot of depth from the surface. If the formation pressure is 4,650 pounds per square inch at 10,000 feet, it is considered normal. See *hydrostatic pressure*; compare *abnormal pressure*.

nozzle *n*: a passageway through jet bits that causes the drilling fluid to be ejected from the bit at high velocity. The jets of mud clear the bottom of the hole. Nozzles come in different sizes that can be interchanged on the bit to adjust the velocity with which the mud exits the bit.

O

oblique slip *n*: slip at an angle between the dip and the strike in a fault plane.

oceanic crust *n*: that part of the earth's outer layer, about 5–7 mi (8–11 km) thick, underlying the oceans. It is composed largely of magnesium, silicon, and oxygen and is thinner and denser than continental crust. Because oceanic crust is created at the midocean ridges and subducted in oceanic trenches where the earth's tectonic plates collide, the oldest oceanic crust is only 200 million years old. Compare *continental crust*.

octane *n*: a paraffinic hydrocarbon, C_8H_{18}, that is a liquid at atmospheric conditions and that has a boiling point of 258° F (at 14.7 pounds per square inch).

octane rating *n*: a classification of gasoline according to its antiknock qualities. The higher the octane number, or rating, the greater the antiknock qualities of the gasoline.

offset roller cone bit *n*: a roller cone bit each cone of which displays offset from a center point when a line is extended through the middle of each cone. Offset roller cone bits are usually employed to drill soft formations, because the offset causes the teeth on the cone to gouge and scrape the formation. Gouging and scraping action is required to penetrate soft formations.

offshore *n*: that geographic area that lies seaward of the coastline. In general, the term "coastline" means the line of ordinary low water along that portion of the coast that is in direct contact with the open sea or the line marking the seaward limit of inland waters.

offshore drilling *n*: drilling for oil or gas in an ocean, gulf, or sea, usually on the outer continental shelf. A drilling unit for offshore operations may be a mobile floating vessel with a ship or barge hull, a semisubmersible or submersible base, a self-propelled or towed structure with jacking legs (jackup drilling rig), or a permanent structure used as a production platform when drilling is completed. In general, wildcat wells are drilled from mobile floating vessels or from jackups, while development wells are drilled from platforms or jackups.

oil *n*: a simple or complex liquid mixture of hydrocarbons, which can be refined to yield gasoline, kerosene, diesel fuel, and various other products.

oilfield *n*: the surface area overlying an oil reservoir or reservoirs. The term usually includes not only the surface area, but also the reservoir, the wells, and the production equipment.

oil patch *n*: (slang) the oilfield.

oil pool *n*: a loose term for an underground reservoir where oil occurs. Oil is actually found in the pores of rocks, not in a pool.

oil sand *n*: 1. a sandstone that yields oil. 2. (by extension) any reservoir that yields oil, whether or not it is sandstone.

oil saturation *n*: the degree to which a formation is saturated by oil, expressed as a percentage of pore volume.

oil seep *n*: a surface location where oil appears, the oil having permeated its subsurface boundaries and accumulated in small pools or rivulets. Also called oil spring.

oil shale *n*: a shale containing hydrocarbons that cannot be recovered by an ordinary oilwell but can that be extracted by mining and processing. The cost of mining and treatment of oil shale is generally too great to compete with the cost of oilwell drilling.

oil-water contact *n*: the plane (typically a zone several feet thick) at which the bottom of the oil zone contacts the top of the water zone in a reservoir, i.e., the oil-water interface.

oilwell *n*: a well from which oil is obtained.

oil-wet reservoir *n*: a hydrocarbon reservoir in which the grains of rock are coated not with water but with oil (occurs only rarely).

oil window *n*: see *petroleum window*.

oil zone *n*: a formation or horizon of a well from which oil may be produced. The oil zone is usually immediately under the gas zone and on top of the water zone if all three fluids are present and segregated.

oolite *n*: rock composed of an accumulation of ooliths.

oolith *n*: an ovoid, sandlike particle that is formed when calcite accretes on a smaller particle.

oolitic *adj*: composed mostly of oolite.

open *adj*: 1. of a wellbore, having no casing. 2. of a hole, having no drill pipe or tubing suspended in it.

open formation *n*: a petroleum-bearing rock with good porosity and permeability.

open hole *n*: 1. any wellbore in which casing has not been set. 2. open or cased hole in which no drill pipe or tubing is suspended. 3. the portion of the wellbore that has no casing.

open-hole completion *n*: a method of preparing a well for production in which no production casing or liner is set opposite the producing formation. Reservoir fluids flow unrestricted into the open wellbore. An open-hole completion has limited use in rather special situations. Also called a barefoot completion.

optimization *n*: the manner of planning and drilling a well so that the most usable hole will be drilled for the least money.

Ordovician *adj*: of or relating to the geologic period from approximately 500 million to 430 million years ago, during the early part of the Paleozoic era, or relating to the rocks formed during this period.

organic compounds *n pl*: chemical compounds that contain carbon atoms, either in straight chains or in rings, and hydrogen atoms. They may also contain oxygen, nitrogen, or other atoms.

organic rock *n*: rock materials produced by plant or animal life (coal, petroleum, limestone, and so on).

organic theory *n*: an explanation of the origin of petroleum that holds that the hydrogen and the carbon that make up petroleum come from land and sea plants and animals. The theory further holds that more of this organic material comes from very tiny swamp and sea creatures than from larger land creatures.

outcrop *n*: part of a formation exposed at the earth's surface. *v*: to appear on the earth's surface (as a rock).

outcrop map *n*: horizontal representation of the rock types at the surface or just beneath the layer of soil.

outer continental shelf (OCS) *n*: the land seaward from areas subject to state mineral ownership to a depth of roughly 8,000 feet, beyond which mineral exploration and development are not, at present, feasible. Boundaries of the OCS are set by law. For example, Louisiana owns 3 miles seaward from the shoreline. In general, the term is used to describe federally controlled areas. Compare *continental shelf*.

outwash *n*: sediment deposited by meltwater streams beyond an active glacier.

overbalance *n*: the extent to which the hydrostatic pressure of the mud column exceeds formation pressure.

overburden *n*: the strata of rock that overlie the stratum of interest in drilling.

overburden pressure *n*: the pressure exerted by the rock strata on a formation of interest. It is usually considered to be about 1.0 pounds per square inch per foot. See *geostatic pressure*.

overthrust belt *n*: any region of the earth's crust characterized by large-scale overthrust faulting.

overthrust fault *n*: a low-dip angle (nearly horizontal) reverse fault along which a large displacement has occurred. Some overthrusts, such as many of those in the Rocky Mountain Overthrust Belt, represent slippages of many miles.

overturned fold *n*: a rock fold that has become slanted to one side so that the layers on one side appear to occur in reverse order (younger layers beneath older)

oxidation *n*: 1. the process of burning. 2. a chemical reaction with oxygen. 3. any chemical reaction in which a compound loses electrons and gains a more positive charge. Compare *reduction*.

oxide *n*: a chemical compound in which oxygen is joined with a metal or a nonmetal.

oxidize *v*: 1. to combine chemically with oxygen. 2. to burn. 3. to remove one or more electrons from an atom, ion, or molecule. See *oxidation*.

oxygenate *v*: supply with oxygen.

oxygenation *n*: the act of supplying with oxygen.

P

paleogeography *n*: geography of a specified geologic past.

paleontology *n*: the science that concerns the life of past geologic periods, especially fossil forms and the chronology of the earth.

Paleozoic era *n*: a span of time from 600 million to 230 million years ago during which a great diversity of life forms developed.

panel diagram *n*: a diagram of a portion of the crust in which a series of cross sections are joined and viewed obliquely from above, to give a three-dimensional view. It is useful in showing how formation structure and stratigraphic thickness vary both horizontally and vertically. Also called fence diagram because it looks like fences joined end-to-end in several directions. Compare *block diagram*.

Pangaea *n*: the supercontinent comprising all of the principal continental masses near the beginning of the Mesozoic era.

passive margin *n*: an area that develops when a growing ocean basin causes continents to drift apart.

pay zone *n*: the producing formation, often one that is not even sandstone. Also called pay, pay sand, and producing zone.

PCD *abbr*: polycrystalline diamond.

PCD bit *n*: a special type of diamond drilling bit that does not use roller cones. Instead, polycrystalline diamond inserts are embedded into a matrix on the bit. PCD bits are often used to drill very hard, abrasive formations, but also find use in drilling medium and soft formations.

penetration rate *n*: see *rate of penetration*.

Pennsylvanian period *n*: a geologic time period in the Paleozoic era, from 320 to 280 million years ago. Also, the latter part of the Carboniferous period. It was named for the outcrops of coal in Pennsylvania.

pentane *n*: a liquid hydrocarbon of the paraffin series, C_5H_{12}.

perforate *v*: to pierce the casing wall and cement of a wellbore to provide holes through which formation fluids may enter or to provide holes in the casing so that materials may be introduced into the annulus between the casing and the wall of the borehole. Perforating is accomplished by lowering into the well a perforating gun, or perforator, which fires electrically detonated bullets or shaped charges.

perforated completion *n*: 1. a well completion method in which the producing zone or zones are cased through, cemented, and perforated to allow fluid flow into the wellbore. 2. a well completed by this method.

perforation *n*: a hole made in the casing, cement, and formation through which formation fluids enter a wellbore. Usually several perforations are made at a time.

permeability *n*: 1. a measure of the ease with which a fluid flows through the connecting pore spaces of rock or cement. The unit of measurement is the millidarcy. 2. fluid conductivity of a porous medium. 3. ability of a fluid to flow within the interconnected pore network of a porous medium. See *absolute permeability, effective permeability, relative permeability*.

permeability barrier *n*: a hindrance to movement of fluids within the formation rock of a reservoir. Permeability barriers include obvious problems, such as shale lenses and calcite or clay deposits, and less obvious ones, such as porosity changes.

permeable *adj*: allowing the passage of fluid. See *permeability*.

Permian Basin *n*: a prolific oil- and gas-producing area in West Texas and eastern New Mexico. So called because many of the rocks from which the oil and gas are withdrawn are of Permian age.

Permian period *n*: the last geologic time period in the Paleozoic era, 280 million to 225 million years ago.

petroleum *n*: a substance occurring naturally in the earth in solid, liquid, or gaseous state and composed mainly of mixtures of chemical compounds of carbon and hydrogen, with or without other nonmetallic elements such as sulfur, oxygen, and nitrogen. In some cases, especially in the measurement of oil and gas, petroleum refers only to oil—a liquid hydrocarbon—and does not include natural gas or gas liquids such as propane and butane. The API Measurement Coordination Department prefers that petroleum mean crude oil and not natural gas or gas liquids.

petroleum geology *n*: the study of oil- and gas-bearing rock formation. It deals with the origin, occurrence, movement, and accumulation of hydrocarbon fuels.

petroleum rock *n*: sandstone, limestone, dolomite, fractured shale, and other porous rock formations where accumulations of oil and gas may be found.

petroleum window *n*: the conditions of temperature and pressure under which petroleum will form. Also called oil window.

petroliferous *adj*: containing petroleum (said of rocks).

petrology *n*: a branch of geology dealing with the origin, occurrence, structure, and history of rocks, principally igneous and metamorphic rocks. Compare *lithology*.

photogeology *n*: the practice of examining aerial and satellite photography of the earth to identify and assess the significance of surface features.

piercement dome *n*: see *diapir*.

pinch-out *n*: an oil-bearing stratum that forms a trap for oil and gas by narrowing and tapering off within an impervious formation.

plate tectonics *n pl*: movement of great crustal plates of the earth on slow currents in the plastic mantle, with large blocks of continental crust imbedded in and carried along on thinner plates of oceanic crust (like boxes on a conveyor belt). Today geologists believe that the earth's crust is divided into six major plates and several smaller ones atop some of which the continents are carried away from a system of midocean ridges and toward another system of deep-sea trenches. The theory of plate tectonics explains most of the mysteries that confounded earlier geologists. Compare *continental drift*.

playa *n*: the flat bottom of an undrained desert basin that at times becomes a shallow lake. Also called sebkha or sabkha.

Pleistocene *adj*: 1. of or relating to the geologic epoch from about 2.5 million to 10 thousand years ago, the first part of the Quaternary period of the Cenozoic era, sometimes called the Ice Age, which extended from the end of the Tertiary period until the last retreat of the northern continental ice sheets. 2. of or relating to the rocks or sediments formed during this epoch.

pluton *n*: a large subterranean body of igneous rock.

point bar *n*: 1. an accumulation of sediment on the inside of a river bend, usually consisting of sand, silt, and clay. 2. a stratigraphic trap for petroleum formed by the burial and lithification of such a sedimentary deposit.

pool *n*: a reservoir or group of reservoirs. The term is a misnomer in that hydrocarbons seldom exist in pools, but, rather, in the pores of rock. *v*: to combine small or irregular tracts into a unit large enough to meet state spacing regulations for drilling.

pore *n*: an opening or space within a rock or mass of rocks, usually small and often filled with some fluid (water, oil, gas, or all three). Compare *vug*.

pore pressure *n*: see *formation pressure*.

pore throats *n pl*: connections between pores.

pore volume (PV) *n*: the total volume of pore space in a reservoir formation.

porosimeter *n*: a device used to determine porosity of a formation by measuring a known volume of gas at a known pressure compressed into a core sample. The gas is measured at atmospheric pressure and at an elevated pressure. These measurements are used to calculate gain volume and pore volume.

porosity *n*: 1. the condition of being porous (such as a rock formation). 2. the ratio of the volume of empty space to the volume of solid rock in a formation, indicating how much fluid a rock can hold. See *absolute porosity, effective porosity, pore*.

porous *adj*: having pores, or tiny openings, as in rock.

pounds per gallon (ppg) *n*: a measure of the density of a fluid (such as drilling mud).

ppg *abbr*: pounds per gallon.

ppm *abbr*: parts per million.

Precambrian era *n*: a span of 4 billion years from the earth's beginning until 600 million years ago, during which the earth was devoid of all but the most primitive life forms.

precipitate *n*: a substance, usually a solid, that separates from a fluid because of a chemical or physical change in the fluid. *v*: to separate in this manner.

pressure *n*: the force that a fluid (liquid or gas) exerts uniformly in all directions within a vessel, pipe, hole in the ground, and so forth, such as that exerted against the inner wall of a tank or that exerted on the bottom of the wellbore by a fluid. Pressure is expressed in terms of force exerted per unit of area, as pounds per square inch, or in kilopascals.

pressure depletion *n*: the method of producing a gas reservoir that is not associated with a water drive. Gas is removed and reservoir pressure declines until all the recoverable gas has been expelled.

pressure differential *n*: see *differential pressure*.

pressure gradient *n*: a scale of pressure differences in which there is a uniform variation of pressure from point to point. For example, the pressure gradient of a column of water is about 0.433 pounds per square inch per foot (9.794 kilopascals per metre) of vertical elevation. The normal pressure gradient in a formation is equivalent to the pressure exerted at any given depth by a column of 10% salt water extending from that depth to the surface (0.465 pounds per square inch per foot or 10.518 kilopascals per metre).

pressure drop *n*: the differential pressure in the flowing liquid stream (which will vary with flow rate) between the inlet and outlet of a meter, flow straightener, valve, strainer, length of pipe, etc.

pressure sink *n*: a condition in which the pressure at the wellbore is less than the reservoir pressure. Flow of oil from the reservoir to the wellbore occurs because of this pressure differential.

primary migration *n*: movement of hydrocarbons out of source rock into reservoir rock. Compare *secondary migration*.

primary porosity *n*: natural porosity in petroleum reservoir sand or rocks, i.e., the porosity developed during the original sedimentation process. Formations having this property (such as sand) are usually granular.

producing horizon *n*: see *pay zone*.

producing interval *n*: see *pay zone*.

producing zone *n*: the zone or formation from which oil or gas is produced. See *pay zone*.

production *n*: 1. the phase of the petroleum industry that deals with bringing the well fluids to the surface and separating them and with storing, gauging, and otherwise preparing the product for the pipeline. 2. the amount of oil or gas produced in a given period.

progradation *n*: the seaward buildup of a beach, delta, or fan by nearshore deposition of sediments by a river, by waves, or by longshore currents.

propane *n*: a paraffinic hydrocarbon (C_3H_8) that is a gas at ordinary atmospheric conditions but is easily liquefied under pressure. It is a constituent of liquefied petroleum gas.

Q

quartz *n*: a hard mineral composed of silicon dioxide (silica), a common component in igneous, metamorphic, and sedimentary rocks.

quartzite *n*: a compact granular rock composed of quartz and derived from sandstone by metamorphism.

R

radial flow *n*: the flow pattern of fluids flowing into a wellbore from the surrounding drainage area.

radioactivity *n*: the property possessed by some substances (such as radium, uranium, and thorium) of releasing alpha particles, beta particles, or gamma particles as the substance spontaneously disintegrates.

radioactivity log *n*: a record of the natural or induced radioactive characteristics of subsurface formations. A radioactivity log, also known as a radiation log or a nuclear log, normally consists of two recorded curves: a gamma ray curve and a neutron curve. Both help to determine the types of rocks in the formation and the types of fluids contained in the rocks. The two logs may be run simultaneously in conjunction with a collar locator in a cased or uncased hole.

rate of penetration (ROP) *n*: a measure of the speed at which the bit drills into formations, usually expressed in feet or metres per hour, or minutes per foot or metre.

reactive solids content *n*: the amount of water-absorbent material in the drilling fluid.

recompletion *n*: after the initial completion of a well, the action and techniques of reentering the well and redoing or repairing the original completion to restore the well's productivity.

record section *n*: a cross section of the earth generated by computer from tapes that have recorded the sound vibrations reflected during seismic exploration. Expert interpretation can reveal what may be a trap for petroleum. Also called seismic section.

recovery *n*: the total volume of hydrocarbons that has been or is anticipated to be produced from a well or field.

recovery efficiency *n*: the recoverable amount of original or residual hydrocarbons in place in a reservoir, expressed as a percentage of total hydrocarbons in place. Also called recovery factor.

recovery factor *n*: see *recovery efficiency*.

recumbent fold *n*: a fold of rock in which the axial plane of an overturned fold has become horizontal or nearly so.

reduction *n*: 1. a chemical reaction in which oxygen is removed from a molecule. 2. any chemical reaction in which one or more electrons is added to an atom, ion, or molecule. Compare *oxidation*.

reef *n*: 1. a type of reservoir trap composed of rock (usually limestone) formed from the shells or skeletons of marine animals. 2. a buried coral or other reef from which hydrocarbons may be withdrawn.

reflection *n*: the return of waves or particles from surfaces on which they are incident.

refraction *n*: deflection from a straight path undergone by a light ray or energy wave in passing from one medium to another in which the wave velocity is different, such as the bending of light rays when passing from air into water.

regional metamorphism *n*: a type of metamorphism that occurs in bodies of rock that have been deeply buried or greatly deformed by tectonic changes.

relative density *n*: 1. the ratio of the weight of a given volume of a substance at a given temperature to the weight of an equal volume of a standard substance at the same temperature. For example, if 1 cubic inch of water at 39° F weighs 1 unit and 1 cubic inch of another solid or liquid at 39° F weighs 0.95 unit, then the relative density of the substance is 0.95. In determining the relative density of gases, the comparison is made with the standard of air or hydrogen. 2. the ratio of the mass of a given volume of a substance to the mass of a like volume of a standard substance, such as water or air.

relative permeability *n*: the ratio of effective permeability to absolute permeability. The relative permeability of rock to a single fluid is 1.0 when only that fluid is present, and 0.0 when the presence of another fluid prevents all flow of the given fluid. Compare *absolute permeability, effective permeability*.

reservoir *n*: a subsurface, porous, permeable rock body in which oil and/or gas has accumulated. Most reservoir rocks are limestones, dolomites, sandstones, or a combination. The three basic types of hydrocarbon reservoirs are oil, gas, and condensate. An oil reservoir generally contains three fluids—gas, oil, and water—with oil the dominant product. In the typical oil reservoir, these fluids become vertically segregated because of their different densities. Gas, the lightest, occupies the upper part of the reservoir rocks; water, the lower part; and oil, the intermediate section. In addition to its occurrence as a cap or in solution, gas may accumulate independently of the oil; if so, the reservoir is called a gas reservoir. Associated with the gas, in most instances, are salt water and some oil. In a condensate reservoir, the hydrocarbons may exist as a gas, but, when brought to the surface, some of the heavier ones condense to a liquid.

reservoir drive *n*: the process in which reservoir fluids are caused to flow out of the reservoir rock and into a wellbore by natural energy. Gas drive depends on the fact that, as the reservoir is produced, pressure is reduced, allowing the gas to expand and provide the principal driving energy. Water drive reservoirs depend on water and rock expansion to force the hydrocarbons out of the reservoir and into the wellbore. Also called natural drive energy.

reservoir pressure *n*: the average pressure within the reservoir at any given time. Determination of this value is best made by bottomhole pressure measurements with adequate shut-in time. If a shut-in period long enough for the reservoir pressure to stabilize is impractical, then various techniques of analysis by pressure buildup or drawdown tests are available to determine static reservoir pressure.

reservoir rock *n*: a permeable rock that may contain oil or gas in appreciable quantity and through which petroleum may migrate.

resistivity *n*: the electrical resistance offered to the passage of current; the opposite of conductivity.

resistivity log *n*: a record of the resistivity of a formation. Usually obtained when an electric log is run. The mineral content of subsurface water allows it to conduct electricity. Rock, oil, and gas are poor conductors. Resistivity measurements can be correlated to formation lithology, porosity, permeability, and saturation and are very useful in formation evaluation. See *electric well log.*

reverse fault *n*: a dip-slip fault along which the hanging wall has moved upward relative to the footwall. Also called a thrust fault.

rift zone *n*: the zone along which crustal plates separate because of slowly diverging convection currents in the semisolid, deformable mantle. As the rift widens and the land masses on both sides move apart, new oceanic crust is formed. Since the Mid-Atlantic rift zone opened about 200 million years ago, North and South America have been moving away from Europe and Africa at a rate of 1¼ inches a year.

rock *n*: a hardened aggregate of different minerals. Rocks are divided into three groups on the basis of their mode of origin: igneous, metamorphic, and sedimentary.

rock bit *n*: see *roller cone bit.*

rock cycle *n*: the possible sequences of events, all interrelated, by which rocks may be formed, changed, destroyed, or transformed into other types of rock. The events include formation from magma, erosion, sedimentation, and metamorphism.

rock stratigraphic unit *n*: a distinctive body of rock that can be identified by its lithologic or structural features regardless of its fossils or time boundaries; commonly, a formation. Compare *time stratigraphic unit.*

rock texture *n*: all of the properties relating to the grain-to-grain relationships of a rock. Textural properties include chemical composition, grain shape and roundness, grain size and sorting, grain orientation, porosity, and permeability. See *clastic texture, crystalline texture.*

roller cone bit *n*: a drilling bit made of two, three, or four cones, or cutters, that are mounted on extremely rugged bearings. The surface of each cone is made of rows of steel teeth or rows of tungsten carbide inserts. Also called rock bits.

rollover anticline *n*: an anticline formed when the dip of a growth fault approaches the horizontal at depth and deposition is faster on the downthrown side, which tends to "roll over" or curl downward.

rollover fault *n*: see *growth fault.*

rotary speed *n*: the speed, measured in rpm, at which the rotary table is operated.

rpm *abbr*: revolutions per minute.

S

sabkha *n*: see *playa.*

salt *n*: 1. a compound that is formed (along with water) by the reaction of an acid with a base. A common salt (table salt) is sodium chloride, NaCl, derived by combining hydrochloric acid, HCl, with sodium hydroxide, NaOH. The result is sodium chloride and water, H_2O. 2. rock salt, or halite.

salt dome *n*: a dome that is caused by an intrusion of rock salt into overlying sediments. A piercement salt dome is one that has been pushed up so that it penetrates the overlying sediments, leaving them truncated. The formations above the salt plug are usually arched so that they dip in all directions away from the center of the dome, thus frequently forming traps for petroleum accumulations. See *diapir.*

salt water *n*: a water than contains a large quantity of salt, i.e., brine.

sand *n*: 1. an abrasive material composed of small quartz grains formed from the disintegration of preexisting rocks. Sand consists of particles less than 2 millimetres and greater than 1⁄16 millimetre in diameter. 2. sandstone.

sand lens *n*: see *lens.*

sandstone *n*: a sedimentary rock composed of individual mineral grains of rock fragments between 1⁄16 and 2 millimetres in diameter and cemented together by silica, calcite, iron oxide, and so forth. Sandstone is commonly porous and permeable and therefore a likely type of rock in which to find a petroleum reservoir.

saturation *n*: a state of being filled or permeated to capacity. Sometimes used to mean the degree or percentage of saturation (e.g., the saturation of the pore space in a formation or the saturation of gas in a liquid, both in reality meaning the extent of saturation).

scarp *n*: an extended cliff or steep slope, produced by erosion or faulting, that separates two level or gently sloping areas, i.e., an escarpment.

sealing fault *n*: a fault that contains material of low permeability, such as gouge.

sebkha *n*: see *playa.*

secondary migration *n*: movement of hydrocarbons that have originated elsewhere through porous, permeable reservoir rock to a location where further migration is impossible (i.e., a trap). Compare *primary migration.*

section *n*: vertically arrayed geologic data; that is, a diagram showing rock strata or other geologic information as though seen in a vertical slice of the crust. Compare *map.*

sediment *n*: solid material, either mineral or organic, that is suspended in, transported by, or moved from its site of origin by flowing air, water, or ice, and has come to rest on the earth's crust either above or below sea level. Sedimentary rocks are composed of such sediments that have been buried, compacted, cemented, and otherwise transformed into rock.

sedimentary basin *n*: a low place where flow energy is dissipated, allowing sediment to settle out of the transporting medium and accumulate.

sedimentary rock *n*: a rock composed of materials that were transported to their present position by wind, water, or ice. Sandstone, shale, and limestone are sedimentary rocks.

sedimentation *n*: the process of deposition of layers of clastic particles or minerals that settle out of wind, water, or other transporting media.

sedimentology *n*: the science dealing with the description, classification, and interpretation of sediments and sedimentary rock.

seep *n*: the surface appearance of oil or gas that results naturally when a reservoir rock becomes exposed to the surface, thus allowing oil or gas to flow out of fissures in the rock.

segregation drive *n*: see *gravity drainage.*

seismic *adj*: of or relating to an earthquake or earth vibration, including those artificially induced.

seismic data *n*: detailed information obtained from earth vibration produced naturally or artificially (as in geophysical prospecting).

seismic method *n*: a method of geophysical prospecting using the generation, reflection, refraction detection, and analysis of sound waves in the earth.

seismic profile *n*: see *record section*

seismic section *n*: see *record section.*

seismic survey *n*: an exploration method in which strong low-frequency sound waves are generated on the surface or in the water to find subsurface rock structures that may contain hydrocarbons. The sound waves travel through the layers of the earth's crust; however, at formation boundaries some of the waves are reflected back to the surface where sensitive detectors pick them up. Reflections from shallow formations arrive at the surface sooner than reflections from deep formations, and since the reflections are recorded, a record of the depth and configuration of the various formations can be generated. Interpretation of the record can reveal possible hydrocarbon-bearing formations.

seismic wave *n*: shock wave generated by an earthquake or explosion.

seismograph *n*: a device that detects vibrations in the earth. It is used in studying the earth's interior and in prospecting for probable oil-bearing structures. Vibrations are created by discharging explosives in shallow boreholes, by striking the surface with a heavy blow, or by vibrating a heavy plate in contact with the ground. The type and velocity of the vibrations as recorded by the seismograph indicate the general characteristics of the section of earth through which the vibrations pass.

seismology *n*: the study of earth vibrations.

self-potential (SP) *n*: see *spontaneous potential.*

shale *n*: a fine-grained sedimentary rock composed mostly of consolidated clay or mud. Shale is the most frequently occurring sedimentary rock.

shale oil *n*: see *oil shale.*

shallow gas *n*: natural gas deposit located near enough to the surface that conductor or surface hole will penetrate the gas-bearing formations. Shallow gas is potentially dangerous because, if encountered while drilling, the well usually cannot be shut in to control it. Instead, the flow of gas must be diverted.

shoestring sand *n*: a narrow, often sinuous, sand deposit, usually a buried sandbar or filled channel.

shoot *v*: in seismographic work, to discharge explosives to create vibrations in the earth's crust. See *seismograph.*

shoreface *n*: that part of the seashore seaward of the low-tide mark that is affected by wave action.

shut in *v*: 1. to close the valves on a well so that it stops producing. 2. to close in a well in which a kick has occurred.

shut-in *adj*: shut off to prevent flow. Said of a well, plant, pump, and so forth, when valves are closed at both inlet and outlet.

silica *n*: a mineral that has the chemical formula SiO_2 (silicon dioxide). It is relatively hard and insoluble. Quartz is a form of silica, but usually contains impurities that give it color.

silicate *n*: crystalline compound composed largely of silicon in chemical combination with aluminum, magnesium, oxygen, and other common elements.

siliceous *adj*: containing abundant silica, or silicon dioxide (SiO_2).

siltstone *n*: a fine grained, shalelike sedimentary rock composed mostly of particles 1/16 to 1/256 millimetre in diameter; intermediate in texture between sandstone and shale.

slate *n*: a metamorphic rock formed when shale becomes buried deeply. The heat and pressure fuse individual mineral grains into slate.

slip plane *n*: adjacent rock surfaces along which differential movement takes place, usually along a fault or fracture. Analogous to surfaces between playing cards.

sloughing (pronounced "sluffing") *n*: see *caving.*

sloughing hole *n*: a condition wherein shale that has absorbed water from the drilling fluid expands, sloughs off, and falls downhole. A sloughing hole can jam the drill string and block circulation.

solid *n*: one of the three physical states of matter. A solid is somewhat rigid and bounds itself internally in all dimensions; therefore, it does not require a container to retain its shape.

solute *n*: a substance that is dissolved in another (the solvent).

solution *n*: a single, homogeneous liquid, solid, or gas phase that is a mixture in which the components (liquid, gas, solid, or combinations thereof) are uniformly distributed throughout the mixture. In a solution, the dissolved substance is called the solute; the substance in which the solute is dissolved is called the solvent.

solution channel *n*: a channel formed by solution action; most commonly, a natural passage dissolved through limestone by the action of a weak solution of carbonic acid (carbon dioxide dissolved in groundwater).

solution gas *n*: lighter hydrocarbons that exist as a liquid under reservoir conditions but that effervesce as gas when pressure is released during production.

solution-gas drive *n*: a source of natural reservoir energy in which the solution gas coming out of the oil expands to force the oil into the wellbore.

solution gas-oil ratio *n*: see *gas-oil ratio.*

solvent *n*: a substance, usually liquid, in which another substance (the solute) dissolves.

sonde *n*: a logging tool assembly, especially the device in the logging assembly that senses and transmits formation data.

sonic log *n*: a type of acoustic log that records the travel time of sounds through objects, cement, or formation rocks. Often used to determine whether voids exist in the cement behind the casing in a wellbore.

sonic logging *n*: see *acoustic well logging.*

sorting *n*: a dynamic process by which different-sized sediments are separated from one another and deposited in different locations or layers as relatively uniform deposits. For instance, a high-energy mountain stream may leave only large cobbles and gravel in one location and carry finer sediments downstream to accumulate in a lower-energy environment, such as a lake.

source rock *n*: rock within which oil or gas is generated from organic materials.

specific gravity *n*: see *relative density.*

spill point *n*: the level at which trapped oil or gas can begin "spilling" upward and out of the trap.

SP log *n*: see *spontaneous potential log.*

spontaneous potential *n*: one of the natural electrical characteristics exhibited by a formation as measured by a logging tool lowered into the wellbore. Also called self-potential or SP.

spontaneous potential curve *n*: a measurement of the electrical currents that occur in the wellbore when fluids of different salinities are in contact. The SP curve is usually recorded in holes drilled with freshwater-base drilling fluids. It is one of the curves on an electric well log. Also called self-potential curve.

spontaneous potential log *n*: a record of a spontaneous potential curve.

sq ft (ft^2) *abbr*: square feet.

sq in ($in.^2$) *abbr*: square inches.

stimulation *n*: any process undertaken to enlarge old channels or to create new ones in the producing formation of a well (e.g., acidizing or formation fracturing).

strata *n pl*: distinct, usually parallel, and originally horizontal beds of rock. An individual bed is a stratum.

stratification *n*: the natural layering or lamination characteristic of sediments and sedimentary rocks.

stratigraphic test *n*: a borehole drilled primarily to gather information on rock types and sequence. Often called a strat test.

stratigraphic trap *n*: a petroleum trap formed principally by a lateral updip discontinuity (such as an unconformity) or decrease in permeability of a reservoir formation. Examples are lenses, reefs, and bar sands. Compare *structural trap.*

stratigraphic unit *n*: unit consisting of stratified, mainly sedimentary, rocks grouped for description, mapping, correlation, or the like.

stratigraphy *n*: a branch of geology concerned with the study of the origin, composition, distribution, and succession of rock strata.

strat test *n*: common name for stratigraphic test.

stratum *n*: singular of strata. A distinct, generally parallel bed of rock.

strike *n*: the horizontal direction of a formation bed or fault plane as measured at a right angle to the dip.

strike slip *n*: horizontal displacement along a fault plane. The San Andreas fault in California is a strike-slip, or lateral, fault.

structural trap *n*: a petroleum trap that is principally the result of deformation (such as folding or faulting) of the reservoir formation. Compare *stratigraphic trap.*

structure *n*: a geological formation of interest to drillers. For example, if a particular well is on the edge of a structure, the wellbore has penetrated the reservoir (structure) near its periphery.

structure contour map *n*: horizontal representation of elevations of a subsurface rock layer or structure.

structure map *n*: a map of critical horizons—subsurface oil-producing zones—in a given area.

stuck pipe *n*: drill pipe, drill collars, casing, or tubing that has inadvertently become immovable in the hole. Sticking may occur when drilling is in progress, when casing is being run in the hole, or when the drill pipe is being hoisted.

subduction zone *n*: a deep trench formed in the ocean floor along the line of convergence of oceanic crust with other oceanic or continental crust when one plate (always oceanic) dives beneath the other. The plate that descends into the hot mantle is partially melted. Magma rises through fissures in the heavier, unmelted crust above, creating a line of plutons and volcanoes that eventually form an island arc parallel to the trench.

subgeologic map *n*: a map of the formations directly above an unconformity. Also called a worm's-eye map.

superposition *n*: 1. the order in which sedimentary layers are deposited, with the oldest layer on the bottom, the youngest layer on top. 2. the process of sedimentary layering.

suspended load *n*: in a flowing stream of water, the finer sand, silt, and clay that are carried well off the bottom by the turbulence of the water. Compare *bed load, dissolved load.*

suspension *n*: a mixture of small nonsettling particles of solid material within a gaseous or liquid medium.

swabbing *n*: a phenomenon characterized by formation fluids being drawn into the wellbore when the drill stem and bit are pulled up the wellbore fast enough to reduce the hydrostatic pressure of the mud below the bit. If enough formation fluid is swabbed into the hole, a kick can result.

syncline *n*: a trough-shaped configuration of folded rock layers. Compare *anticline.*

T

tectonic *adj*: of or relating to the deformation of the earth's crust, the forces involved in or producing such deformation, and the resulting rock forms.

thrust fault *n*: see *reverse fault.*

thumper *n*: a hydraulically operated hammer used in obtaining a seismograph in oil exploration. It is mounted on a vehicle and, when dropped, creates shock waves in the subsurface formations, which are recorded and interpreted to reveal geological information.

tight formation *n*: a petroleum- or water-bearing formation of relatively low porosity and permeability.

tight sand *n*: sand or sandstone formation with low permeability.

time stratigraphic unit *n*: a layer of rock, with or without facies variations, deposited during a distinct geologic time interval. Compare *rock stratigraphic unit.*

topographic map *n*: horizontal representation of land elevations as a series of contour lines, each line connecting points of equal elevation.

topography *n*: the configuration of a land surface, including its natural and artificial features with their relative positions and elevations.

topset bed *n*: a part of a marine delta that is nearest the shore and that is composed of the heavier, coarser particles carried by the river.

total porosity *n*: see *absolute porosity.*

transform fault *n*: a strike-slip fault caused by relative movement between crustal plates.

trap *n*: a body of permeable oil-bearing rock surrounded or overlain by an impermeable barrier that prevents oil from escaping. The types of traps are structural, stratigraphic, hydrodynamic, or a combination of these.

trilobite *n*: one of a class of extinct Paleozoic arthropods.

trip *n*: the operation of hoisting the drill stem from and returning it to the wellbore. *v*: shortened form of "make a trip."

trip in *v*: to lower the drill stem, tubing, casing, or sucker rods into the wellbore.

trip out *v*: to pull the drill stem out of the hole to change the bit, change from a core barrel to the bit, run electric logs, prepare for a drill stem test, run casing, or for other purposes.

tubing *n*: relatively small-diameter pipe that is run into a well to serve as a conduit for the passage of oil and gas to the surface.

tubing head *n*: a flanged fitting that supports the tubing string, seals off pressure between the casing and the outside of the tubing, and provides a connection that supports the Christmas tree.

tubular goods *n pl*: any kind of pipe. Oilfield tubular goods include tubing, casing, drill pipe, and line pipe. Also called tubulars.

tubulars *n pl*: shortened form of tubular goods.

tungsten carbide *n*: a fine, very hard, gray crystalline powder, a compound of tungsten and carbon. This compound is bonded with cobalt or nickel in cemented carbide compositions and used for cutting tools, abrasives, and dies.

tungsten carbide bit *n*: a type of roller cone bit with inserts made of tungsten carbide. Also called tungsten carbide insert bit.

U

ultimate recovery *n*: total anticipated recovery of oil or gas from a well, lease, or pool.

unconformity *n*: 1. lack of continuity in deposition between rock strata in contact with one another, corresponding to a gap in the stratigraphic record. 2. the surface of contact between rock beds in which there is a discontinuity in the ages of the rocks. See *angular unconformity, disconformity.*

unconsolidated formation *n*: a loosely arranged, apparently unstratified section of rock.

unconsolidated sandstone *n*: a sand formation in which individual grains do not adhere to one another. If an unconsolidated sandstone produces oil or gas, it will produce sand as well if not controlled or corrected.

underbalanced *adj*: 1. of or relating to a condition in which pressure in the wellbore is less than the pressure in the formation. 2. drilled with underweight mud or clear water whose hydrostatic pressure is less than formation pressure.

underground blowout *n*: an uncontrolled flow of gas, salt water, or other fluid out of the wellbore and into another formation that the wellbore has penetrated.

uniformitarianism *n*: the geologic principle that the processes that are at work today on the earth are the same as, or very similar to, the processes that affected the earth in the past. The principle was formulated in 1785 by James Hutton, a Scottish geologist. It became an accepted doctrine with the publication in 1830 of English geologist Charles Lyell's book *Principles of Geology*. Also called gradualism. Compare *catastrophism*.

updip *adj*: higher on the formation dip angle than a particular point.

upstructure *adj*: toward the upper portion of an updip formation.

V

vapor *n*: a substance in the gaseous state that can be liquefied by compression or cooling.

velocity boundary *n*: the plane of contact between two formations or rock layers that transmit seismic waves at different velocities because of differences in density, composition, texture, or other characteristics. Seismic waves tend to be reflected from velocity boundaries, which makes them useful in locating the tops and bottoms of different rock strata.

vertical permeability *n*: the permeability of a pay zone from the bottom to the top of the zone. Compare *horizontal permeability*.

vesicular *adj*: containing small cavities that are visible to the naked eye. A vesicular rock is a rock that contains small cavities.

viscosity *n*: a measure of the resistance of a fluid to flow. Resistance is brought about by the internal friction resulting from the combined effects of cohesion and adhesion. The viscosity of petroleum products is commonly expressed in terms of the time required for a specific volume of the liquid to flow through a capillary tube of a specific size at a given temperature.

viscous *adj*: having a high resistance to flow.

void *n pl*: a cavity in a rock that does not contain solid material but may contain fluid.

volatile *adj*: readily vaporized.

volatility *n*: the tendency of a liquid to assume the gaseous state.

volcanology *n*: the scientific study of volcanoes and volcanic phenomena. Also called vulcanology.

vug *n*: 1. a cavity in a rock. 2. a small cavern, larger than a pore but too small to contain a person. Typically found in limestone subject to groundwater leaching.

vuggy formation *n*: a rock formation that contains openings (vugs). See *vug*.

vuggy porosity *n*: secondary rock porosity formed by the dissolving of the more soluble portions of a rock in waters containing carbonic or other acids.

vugular formation *n*: see *vuggy formation*.

vulcanology *n*: see *volcanology*.

W

wall cake *n*: also called filter cake or mud cake. See *filter cake*.

wall sticking *n*: see *differential sticking*.

wall-stuck pipe *n*: see *differential sticking*.

water coning *n*: the upward encroachment of water into a well caused by pressure drawdown from production.

water drive *n*: the reservoir drive mechanism in which oil is produced by the expansion of the underlying water and rock, which forces the oil into the wellbore. In general, there are two types of water drive: bottom-water drive, in which the oil is totally underlain by water; and edgewater drive, in which only a portion of the oil is in contact with the water.

water drive field *n*: an oilfield in which the primary reservoir drive mechanism is water.

watered-out *adj*: of a well, producing mostly water ("gone to water").

water encroachment *n*: the movement of water into a producing formation as production depletes the formation of oil and gas.

water saturation *n*: the percentage of total pore space in a reservoir occupied by water. It is normally estimated by electric log calculations, although cores specially obtained may be analyzed for this value.

water table *n*: the underground level at which water is found.

water-wet reservoir *n*: a hydrocarbon reservoir whose rock grains are coated with a film of water.

water zone *n*: the portion of an oil or gas reservoir occupied by water, usually the lowest zone in the reservoir.

weathering *n*: the breakdown of large rock masses into smaller pieces by physical and chemical climatological processes.

weight on bit (WOB) *n*: the amount of downward force placed on the bit by the weight of the drill collars.

weight up *v*: to increase the weight or density of drilling fluid by adding weighting material.

well *n*: the hole made by the drilling bit, which can be open, cased, or both. Also called borehole, hole, or wellbore.

wellbore *n*: a borehole; the hole drilled by the bit. A wellbore may have casing in it or it may be open (uncased); or part of it may be cased, and part of it may be open. Also called a borehole or hole.

well completion *n*: 1. the activities and methods of preparing a well for the production of oil and gas or for other purposes, such as injection; the method by which one or more flow paths for hydrocarbons are established between the reservoir and the surface. 2. the system of tubulars, packers, and other tools installed beneath the wellhead in the production casing; that is, the tool assembly that provides the hydrocarbon flow path or paths.

well control *n*: the methods used to control a kick and prevent a well from blowing out. Such techniques include, but are not limited to, keeping the borehole completely filled with drilling mud of the proper weight or density during all operations, exercising reasonable care when tripping pipe out of the hole to prevent swabbing, and keeping careful track of the amount of mud put into the hole to replace the volume of pipe removed from the hole during a trip.

wellhead *n*: the equipment installed at the surface of the wellbore. A wellhead includes such equipment as the casinghead and tubing head. *adj*: pertaining to the wellhead (e.g., wellhead pressure).

well logging *n*: the recording of information about subsurface geologic formations, including records kept by the driller and records of mud and cutting analyses, core analysis, drill stem tests, and electric, acoustic, and radioactivity procedures. See *acoustic log, core analysis, driller's log, drill stem test, electric well log,* and *radioactivity log*.

well stream *n*: the fluid, usually a combination of gas, oil, water, and suspended sediment, that comes out of a reservoir.

wildcat *n*: 1. a well drilled in an area where no oil or gas production exists. 2. (nautical) the geared sheave of a windlass used to pull anchor chain. *v*: to drill wildcat wells.

wildcatter *n*: one who drills wildcat wells.

wild well *n*: a well that has blown out of control and from which oil, water, or gas is escaping with great force to the surface. Also called a gusher.

wireline *n*: a small-diameter metal line used in wireline operations. Also called slick line. Compare *conductor line*.

wireline log *n*: any log that is run on wireline.

wireline workover *n*: a workover performed with wireline lowered into the well through the tubing string. A lubricator is rigged up over the wellhead and wireline tools are inserted through the lubricator under pressure. A variety of wireline tools is available.

work over *v*: to perform one or more of a variety of remedial operations on a producing oilwell to try to increase production. Examples of workover operations are deepening, plugging back, pulling and resetting liners, and squeeze cementing.

workover *n*: the performance of one or more of a variety of remedial operations on a producing oilwell to try to increase production. Examples of workover jobs are deepening, plugging back, pulling and resetting liners, and squeeze cementing.

workover fluid *n*: a special drilling mud used to keep a well under control while it is being worked over. A workover fluid is compounded carefully so that it will not cause formation damage.

workover rig *n*: a portable rig used for working over a well.

worm's-eye map *n*: see *subgeologic map*.

Z

zone *n*: a rock stratum that is different from or distinguished from another stratum (e.g., a pay zone).

zone isolation *n*: the practice of separating producing formations from one another by means of casing, cement, and packers for the purposes of pressure control and maintenance, as well as the prevention of mixing of fluids from separate formations.

zone of lost circulation *n*: a formation that contains holes or cracks large enough to allow cement to flow into the formation instead of up along the annulus outside of the casing.